LIGHT IT UP

LIGHT IT UP

The Marine Eye for Battle
in the War for Iraq

JOHN PETTEGREW

Johns Hopkins University Press
Baltimore

© 2015 Johns Hopkins University Press
All rights reserved. Published 2015
Printed in the United States of America on acid-free paper
9 8 7 6 5 4 3 2 1

Johns Hopkins University Press
2715 North Charles Street
Baltimore, Maryland 21218-4363
www.press.jhu.edu

Library of Congress Cataloging-in-Publication Data

Pettegrew, John, 1959–
 Light it up : the Marine eye for battle in the war for Iraq / John Pettegrew.
   pages cm
 Includes bibliographical references and index.
 ISBN 978-1-4214-1785-1 (hardcover : alk. paper) — ISBN 978-1-4214-1786-8 (electronic) — ISBN 1-4214-1785-5 (hardcover : alk. paper) — ISBN 1-4214-1786-3 (electronic) 1. Iraq War, 2003–2011. 2. Iraq War, 2003–2011—Psychological aspects. 3. Combat—Psychological aspects. 4. United States. Marine Corps.—Iraq. 5. Video games—Social aspects. 6. Computer war games—Social aspects. 7. Marines—United States—Psychology. 8. Optics—Social aspects—History—21st century. 9. Visual communications—United States—Social aspects—History—21st century. 10. Military art and science—United States—Computer simulation methods—Social aspects. 11. Iraq War, 2003–2011—Moral and ethical aspects. 12. Strategic culture—United States. I. Title. II. Title: the Marine eye for battle in the war for Iraq.
 DS79.76.P482 2015
 956.7044′34—dc23   2014049510

A catalog record for this book is available from the British Library.

Special discounts are available for bulk purchases of this book. For more information, please contact Special Sales at 410-516-6936 or specialsales@press.jhu.edu.

Johns Hopkins University Press uses environmentally friendly book materials, including recycled text paper that is composed of at least 30 percent post-consumer waste, whenever possible.

For Tamara

Contents

Preface ix
Acknowledgments xiii

Introduction. Force Projection and the Marine Eye for Battle 1

1 Shock and Awe and Air Power 12
  Network-Centric Warfare, Sensors, and Total Situational Awareness 13
  Achieving Rapid Dominance in Iraq 17
  Kill Boxes, LITENING Pods, and the Third Marine Aircraft Wing 20
  "Keep Your Eyes Out," Fair Fighting, and Memories of Killing 26

2 Of War Porn and Pleasure in Killing 37
  Pornography Is the Theory, and Killing the Practice 41
  Classic Hollywood Combat Films 45
  Marine Moto on YouTube 49
  The Iraq War on Television 57

3 Fallujah, *First to Fight*, and Ludology 65
  Ender's Game and the Rise of Simulation in Military Training, 1995–2005 69
  From Combat Films to Video Games 78
  The Value Added to Military Training 82
  Fighting in the Digitized Streets of Beirut 86

4 Counterinsurgency and "Turning Off the Killing Switch" 96
  Empathy, General Mattis, and the Profound Paradox of Marine Humanitarianism 100
  Haditha, Acute Stress, and the Excesses of Occupying Force 104
  USMC Literary Culture and Warrior Ethos 112
  "Which Way Would You Run?" 118

5 Posthuman Warfighting 127
   Marines in Science Fiction and in Space 131
   The Postmasculinist Marines and New Optics of Combat 136
   The Gladiator Robot and the Critique of Remote Warfare 142

6 Synthetic Visions of War: Conclusion and Epilogue 149
   Biopolitics and the Costs of War 151
   Digital Culture and the Computational Marine 155
   Subjectivity Lives and Dies 167

Notes 177
Essay on Primary Sources 201
Index 203

Preface

During the past twelve years of studying the United States Marine Corp (USMC) in the Iraq War, personal encounters with three individual marines have deepened my understanding of what would become this book's primary focus, the USMC's institutionally, technologically, and culturally derived view toward combat—the Marine eye for battle. My earlier book on post–Civil War gender and violence has the USMC as the Teddy Roosevelt of the American armed forces: imbued with "fighting spirit" to the point of caricature, the Corps seems to require little historical explanation, especially when it comes to how Marine celebration of its fighting and killing prowess helped establish a hypermasculinist warrior class in the late nineteenth- and twentieth-century United States. Marine Corps culture, in all its functional richness, involves a self-conscious revelry in warfighting as patriotic blood sport. Yet its institutional memory and culture also bear an independent streak of skepticism toward the heavy military lifting involved in extending and maintaining U.S. empire. This combination of "marines praying for war" and a doubtful sober-mindedness over war's great costs came through in each of my three marine meetings during the conflict in Iraq.

The first encounter was with General Anthony Zinni (ret.) when he came to speak at Lehigh University in early March 2003, just days before the U.S. invasion of Iraq. Confidently understated in his person, dressed in khaki pants, an open-collar shirt, and blue blazer, Zinni struck me as another middle-aged academic during the prelecture reception and dinner. At the university's auditorium, though, with more than two thousand people in attendance, Zinni commanded authority over the evening's topic, the coming Iraq War, like no international relations or political science professor ever could.

Standing still on an empty stage, with his hands clasped behind his back, Zinni spoke without notes for well over an hour, detailing how the United States was about to make a stupendous strategic mistake, one that would harm national security, tear Iraq into pieces, and bring dangerous instability to the whole Gulf region. Zinni's credibility couldn't have been greater. Chief of the U.S. Central Command (responsible for theater-level operations in the Middle East, North Africa, and Central Asia) from 1997 to 2000, Zinni had retained his top-secret clearances while consulting with the CIA about Iraq and terrorism after September 11, 2001. He had seen the intelligence reports, and nothing suggested weapons of mass destruction (WMDs) in Iraq. He knew nothing was there. "There may be a petri-dish somewhere in Iraq with the beginnings of a biological weapon," I recall him saying, "but I doubt even that. And I can assure you that WMDs are simply not to be found in Iraq." On top of that, he continued, the Pentagon has planned the invasion with a woefully undersized military force. Most importantly, in deciding to invade, the Bush administration doesn't know what it's getting into. Sectarian strife will require U.S. military occupation of a post–Saddam Iraq for years to come, a costly operation in blood and treasure that the American people have no real interest in. It was the most remarkable political moment I've experienced. "I've bled four times for my country," Zinni said in referring to his service in the Vietnam War. "I'm a patriot," he implored. Leaving the auditorium that evening, I remember thinking Zinni made that last statement to explain why he opposed the Iraq War so fervently.

The second encounter came just a few months later in 2003, after the successful Iraq invasion and the toppling of Saddam's regime. It was late August, the beginning of a new school year, when I met Sergeant Colin Keefe (ret.) after the opening lecture in my post-1939 U.S. history survey. He approached me at the lectern, introduced himself, and stated that during the semester he would be respectfully disagreeing with my criticism of the Global War on Terror. Dressed in jeans and a t-shirt, at first glance Keefe looked at home among the Lehigh undergrads. But then he proceeded to tell me that he had just been discharged from the U.S. Marines. Over the past spring and early summer he had been a machine gunner in the invasion of Iraq and the battle over Baghdad. We looked at each other intently, both thinking how strange it was for him to be on campus, in this classroom, so shortly after he had fought in Operation Iraqi Freedom.

I got to know Keefe fairly well over the next few years. He became a history major and spoke with me and to my students about Iraq and the early twenty-first-century U.S. military. With my interest in this book project growing, Keefe agreed to sit down for two lengthy audiotaped interviews. It was only then that I learned of his postwar troubles stemming from the continuing incongruity between civilian life as a college student and the Iraq War in his head. He convinced me that he had gotten through the worst of it. In looking back at his service in the Marines, he insisted that it had been a constructive step for a young upper-middle-class kid who almost flunked out of Lehigh before joining up in 1999. His father, now an investment banker in New York City, had gone through a very similar experience with Brown University, the Army, and the Vietnam War. The Marines had taught Colin focus, discipline, and commitment. The Iraq War wasn't going well, and he still had nightmares from it, but on balance he knew he had gained direction from his service in the Marines.

The third marine meeting happened several years later, in summer 2012, as part of my ongoing oral history project on U.S. military veterans from the Iraq and Afghanistan wars. A Lehigh graduate student, Robert O'Neil, suggested that I speak with his cousin Sergeant Thomas Lowry (ret.), who had done several tours of duty in Iraq and Afghanistan. Robert cautioned me. His cousin was having a very difficult time. In and out of treatment for trauma and stress, having multiple brushes with the law for disorderly conduct, Lowry had no real home or job or hope of finding either one. I made arrangements to meet Lowry a half-hour before the interview so that we could get a sense of each other before taping began. Even though it was a hot summer day, he wore a Marine field jacket with his name above the chest pocket but also adorned with a skull and crossbones on the back among other unofficial symbols and insignias. He expressed enthusiasm about our talk, eagerly showing me a wound on his leg from an explosive device at a Baghdad checkpoint, and demonstrating to me in conversation the film, the fog, the veil that separated his person from the reality of those around him.

Upon starting the interview I quickly learned that life before the Marines was not terribly pleasant or stable for Lowry. With his father having trouble staying in a job, Lowry had moved around a lot during his childhood. High school didn't mean much to him outside of playing football. And so, in 2000, without many choices, and like so many before him, Lowry decided to join the Marines. I discuss Lowry's wartime experience in chapter 4, and my full interview with him can be seen on the Veterans Empathy Project website. Those interested in him might find their own answers to what the Marines meant for Lowry. I'm still trying to balance the damage that he did to others in the name of the U.S. Marines and the Global War on Terror with how that conflict obliterated his person and life, which ended in a 2013 car crash. I do know, though, that Lowry is just as representative of the Marine eye for battle as Keefe or Zinni. This book, while keeping multiplicity of perspective in mind, fills in the ways that the Marine Corps patterned the optics of combat in the Iraq War. Ingenious in facilitating the killing of strangers—face-to-face and from great distance—Marine visual culture is far less suited for mediating peace for one's self or with others.

Acknowledgments

My historical approach to war as a human institution began with my graduate school teacher and friend Paul Boyer (1935–2012). Since war and peace are products of social thought, intellectual history has a special role in their examination. It's a type of critical analysis that should involve the historian's moral imagination—carefully applied and preferably with democratic sensibility. That's where I started with him at Wisconsin. And it's where I ended up: a civic-minded subject position of historical inquiry, modeled by both Paul Boyer and his Madison predecessor, the pragmatist historian Merle Curti.

Long-term intellectual debts include other leading historians of war and the militarization of modern American thought, culture, and society. Michael Sherry's studies of "the shadow of war" in the twentieth-century United States and the rise of strategic air power during World War II have been invaluable to my thinking about, among many other themes, "the attractions and dangers of a dehumanized technology of war: destructive passion disguised as cold science, peril clothed in progress, imperialism masked as ingenuity."[1] Important to my turn to visual culture and war has been John Dower's book on the viscerally determined hatred and atrocity between Japanese and U.S. forces in the Pacific during World War II. Also influential has been Joanna Bourke's study of face-to-face killing in war.

Jumping ahead to the twenty-first century, I've benefited greatly from the dogged and fervent antiwar writing following the September 11 attacks. I'm especially beholden to Tom Engelhardt, Nick Purse, Andrew Bacevitch, and other critics able to traverse the line between academic scholarship and wider political discourse. I've written my book in line with the increasingly rigorous examination of military cultural history. I'm particularly indebted to Beth Bailey and Aaron B. O'Connell for their creative institutional histories of, respectively, the U.S. Army and the U.S. Marines.

In a different vein, the latter parts of my book work from historian Donna Haraway's rich and creative insights regarding questions of social power and the contingency of the species line. Just as her blurring of the human-ape divide contributed to my first book on masculinity, her postulations on cyborgian redemption influenced this work's material on posthuman warfighting.

Thank you to all the veterans who spoke with me about their wartime experience for the Veterans Empathy Project.

I could not have written this close examination of the U.S. Marines in the Iraq War without the cooperation of the Corps itself. Its Division of Public Affairs proved helpful in connecting me with different USMC offices and personnel. I spent considerable time in Quantico, Virginia, at Marine Corps University, where the staffs at the Library of the Marine Corps and at the Marine Corps Archives were generous with their time and skilled in locating material important to this study. Paula Holmes-Eber and other anthropologists at Quantico's Center for Advanced Operational Culture Learning were professionally gracious in sharing their work on counterinsurgency doctrine with me and helping me understand their role within Marine Corps training. Most of all, thank you to the History Division of the United States Marine Corps for its hospitality and source material offered throughout my research trips to Quantico. It made a huge difference. Special thanks to Dr. Nicholas J. Schlosser, formerly of the History Division, for welcoming me to Quantico, sharing his deep knowledge of the Marines and military history with me, and for reading a full draft of my book manuscript.

For my chapter on first-person shooter combat video games, the Destineer Corporation of Plymouth, Minnesota, maker of *Close Combat: First to Fight*, was wonderful in opening its studio and production process to me. President Peter Tamte and designer David Kuykendall were patient and informative in their interviews with me.

My study received many types of institutional support from Lehigh University, including separate research grants from the Office of Research and Graduate Studies, Library and Technology Services, and the Humanities Center. Johanna Brams and her staff at Web and Mobile Services were crucial in helping me capture and put together early presentations using combat video-game clips. The Digital Media Studio's Steve Lichak, Allen Kingsbury, and Jarett Brown have been indispensable to the oral history component of this study, as have Bobby Siegfried, Chris Brockman, and Liz Erwin. Lehigh's Center for Innovations in Teaching and Learning gives a new name to long-standing friends, colleagues, and working relationships. Greg Reihman has offered strong support for this work and related efforts. And from the first time I presented this material on the Marines in the Iraq War until today, Julia Maserjian has been steadfast in focusing her considerable enthusiasm, know-how, and creativity on my project; her expansive expertise, political acumen and sociability, and warm friendship have helped sustain me over the past several years.

The work has been moved along by a number of Lehigh graduate students and research assistants. Thanks very much to Kate Meiman, Jen Smith, Christianne Gadd, Jerry Harnett, and William Ying. Matt Schandler shared his impressive expertise on video-game scholarship in reading an early draft of my chapter on Marine simulation and video gaming.

In addition to presenting parts of this study in different Lehigh venues, I'm grateful for having the opportunity to discuss my work at several other institutions. Nancy Cott's invitation to speak at Harvard University's Charles Warren Center for Studies in American History in 2006 provided a timely forum for developing the project's scope and conceptual themes. Thank you to the 2008 Peace, War and Global Change Seminar at the University of Kansas for its critical attention to my interdisciplinary study; Sheyda Jahanbani and Jonathan Hagel were wonderful hosts in Lawrence, and they remain beloved touchstones for intellectual, familial, and gastronomic endeavors. Thank you also to my former student Michael King and the History and Culture Program at Drew University for inviting me to give the keynote address at the school's 2013 annual cultural studies conference.

I'm indebted to Eric Fair, Steven Pressfield, Evan Reibsome, Roger Stahl, and Bing West for different acts of generosity and support of my work over the past several years. My colleagues in Lehigh's History Department, Humanities Center, and American Studies Program have made up a good, comfortable, and stimulating environment for intellectual work. Thank you to Steve Cutcliffe and John Smith, in particular, for reading a significant portion of my manuscript and providing cogent criticism and collegial encouragement. Saladin Ambar, Barry Kroll, Seth Moglen, Ted Morgan, Ziad Munson, and John Savage listened intently to my talk of marines along the way, while also providing much needed refuge through friendship from fixing on the Global War on Terror.

My parents continue to impress and support me with their love; it's returned more than you know. Helen Keetley and Nick Pettegrew have lived with this project for a good long while. Thanks for your dog walking and dish cleaning. Thanks for your quiet interest in my work. Most of all, though, thanks for your patience. That Tamara Myers joined me in academic life's journey has buoyed me to no end. Her well-informed encouragement at each juncture and her critical reading of chapters amid her own extremely busy work life have been crucial in seeing the project to its conclusion. It's for her love and support and the promise of life together beyond our work to which I dedicate this book.

INTRODUCTION

Force Projection and the Marine Eye for Battle

This work on the U.S. Marines and the early twenty-first-century visual culture of combat stems from its author's opposition to the Iraq War. That needs to be said upfront. What follows connects in subject matter with my earlier book on modern American masculinity and violence and its postscript on how popular war literature and Hollywood combat movies helped create an optically calibrated, battle-ready mindset among the emergent post–World War II warrior class.[1] The relationship between culture and violence in war remains a predominant intellectual interest of mine. But the decision to invade Iraq in the wake of the September 11, 2001, attacks provided the major impetus for this undertaking in contemporary history. The spring 2003 shock and awe invasion caused a still unfolding disaster for Iraq's people, ecology, infrastructure, and institutions. Morally, legally, and strategically unsound, Operation Iraqi Freedom dropped the United States into a $2.3 trillion war of occupation and counterinsurgency responsible for over 36,000 American casualties and virtually countless Iraqi dead and wounded.[2] And the cycle of violence and extremism so amplified in the Iraq War continues well after the withdrawal of U.S. ground forces. ISIS has joined Al Qaeda and other extremist Islamic groups in escalating operations against Muslims and non-Muslims alike. Iraq, Syria, and Yemen are engulfed in civil warfare. The United States and the West continue to suffer attacks on their civilians, while the American military relies on ever more remote weapon systems to take out those it calls terrorists.

While millions upon millions of people from around the world protested the Bush administration's run-up to the invasion, U.S. and Coalition forces organized and unleashed, in the words of Commanding General Tommy Franks, a "fast and final" attack on Iraq: an operation featuring high-tech information and communications systems, devastating air power, and 195,000 ground troops.[3] That number would prove to be too small for securing the country after Saddam Hussein's defeat, but the ground force's offensive capacity was extremely formidable in its mobility, technology, organization, and firepower.

In the beginning of a century that will most likely see automated weapon systems become the primary means of warfighting, the U.S.-led invasion of Iraq still hinged on the brute matter of putting boots on the ground and ordering troops to kill the enemy in front of them. To put a finer point on it, Operation Iraqi Freedom depended on a premeditated spasm of violence largely carried out by young American men with little personal reason to take the lives of those whose land they had invaded. That they could do so with such success and apparent ease is the particular occasion for this study of the United States Marine Corps, whose First Division, along with the U.S. Army's V Corps, made up most of the ground force that invaded and then, with the Second Division and the Army, occupied Iraq.

Despite opposing the Iraq War, I seek neither to judge individual marines morally culpable nor to damn the United States Marine Corps as a flawed institution. Having recognized my own investment in this book's historical moment, I've tried deliberately to move to a more objective plane by addressing its issues from the perspectives of its key subjects. My research has fostered a certain respect for the Corps' idealism, adaptability, and commitment to what it does best. The aim and content of that action are largely destructive and rarely necessary. But that action has come only from decisions made by political figures higher in the chain of command. The Corps has long understood its subordinate role as an all-volunteer colonial force. A display in the Marine Corps Museum in Quantico, Virginia, on early twentieth-century wars includes a quote from Colonel Joseph Alexander: "America was suddenly an imperial nation and the Marines had a lot of Uncle Sam's dirty work to do." The Marines' embrace of aggressiveness as a character trait is excessive and proud. Nothing if not self-realized, the Corps has transformed the idea and feeling of military violence into a signature cultural complex, a heroic warrior tradition with hypermasculinity at its ideological core.[4]

While not dropping masculinity from consideration, this book moves from gender to optics as its primary category of analysis. Gender and sight, of course, are not mutually exclusive. Hypermasculinity is not only an identity source that influences behavior and social relations but also a way of looking at the world—a cognitive and emotionally laden frame of perception. It melds well with other systems of seeing more explicitly fashioned for wartime killing. Critical attention to those systems yields a better understanding of U.S. war making, while empirical evidence from the early twenty-first century provides further reason for shifting focus to the optics of combat. The corpus of

American cultural production from the post–September 11 wars—the memoirs, battle histories, newspaper articles, TV shows, blogs, feature films, YouTube videos, and videogames—are not fixed on masculinity as much as their early twentieth century counterparts. It holds true for the Marines as well as the U.S. Army. And so, I argue, U.S. warfighting in Iraq was motivated less by physical prowess idealized in the male warrior form, and more by acquiring techniques of violence located in the eye and expressed through skills of killing from a distance. Killing the enemy in Iraq rested more in the eye, as a matter of built-in reflex or "trained instinct," than in discursive appeals to upholding one's manhood by killing in combat.[5]

Automated killing in combat runs back to the immediate post–World War II era, when behavioral science attempted to substantiate the psychological blocks against killing in war. S. L. A. Marshall's controversial book *Men against Fire* (1947) is an essential reference point here. Based on four hundred interviews with infantry companies in the Pacific and European theaters, the study documents that only 15 to 20 percent of American troops in World War II shot at a targeted enemy soldier. While Marshall's methodology has been roundly criticized, later behavioral and social scientists have confirmed his claim regarding the nearly wholesale ineffectiveness of combat troops. Military psychologist Dave Grossman's influential work *On Killing: The Psychological Cost of Learning to Kill in War and Society* (1995) synthesizes widely varying historical accounts of soldiers routinely aiming to miss the enemy or not shooting at all. Such "conscientious objection" to taking another life, the book posits, comes from an innate human fear of killing. "Looking another person in the eye," Grossman writes, "making an independent decision to kill him, and watching as he dies due to your action combine to form the single most basic, important, primal, and potentially traumatic occurrence of war." Grossman's study is most helpful in summarizing the U.S. armed services' application of behaviorist psychology after World War II in training recruits to kill: the military replaced stationary bull's-eye targets with "pop-up" man-shaped figures; rather than lying comfortably prone while shooting, the troops, dressed in full combat equipment, would stand for hours in a foxhole or other battle-like conditions; and the target, once hit, provided "immediate feedback by instantly and very satisfyingly dropping backward—just as a living target would."[6] By the end of the century, Grossman reports, the nonfiring rate for U.S. troops in combat was very close to zero.

• • •

The point to take from Marshall and Grossman is not that some 85 percent of the World War II marines fighting in the Pacific failed to shoot at Japanese troops. That figure remains very much in doubt.[7] It is, rather, that the Marines, along with the U.S. military overall, accepted the nonfiring problem as true; proceeded to implement stimulus-response and see-shoot infantry training; and, generally, worked to perfect the willingness and ability of a marine or soldier to shoot and kill another human being. In the broadest terms, reflexive shooting of the enemy and technological advances in weaponry that distanced killer from killed combined to move warfighting toward a systematized application of violence that increasingly depended on optics and seeing to connect with one's target and destroy it. For the Marines, that would mean all the more emphasis on the "marine and his rifle"—whom Army general John Pershing during World War I called the "deadliest weapon in the world"—as the centerpiece of the Corps' role and identity as a highly mobile assault force.

U.S. Marine recruit training, or boot camp, entails twelve weeks (two weeks longer than the Army's basic training) of drill, martial training, and physical and mental testing. Its purposes are many, including introduction to the Corps' distinctive and powerful culture, dissolution of individual identity, and group cohesion.[8] Whether labeled "acculturation," "indoctrination," or "brainwashing," recruits are immersed in the Corps' formal values and ethics as well as the stories and traditions that express the elite and superior status of the Marines among all other military services. Historian Aaron B. O'Connell valuably calls them "narratives of Marine exceptionalism."[9] Boot camp culture is conceptually and linguistically heavy on killing. Bayonet training and pugil sticks are meant to breed aggressiveness. The word *kill* is introduced as synonymous with *aye* and used to acknowledge an order, and more: "Every response was 'Kill,'" as a former marine recruit from 2002 describes:

> Every chant we had, whether it was in line for the chow hall or PT [physical training] was somehow involved with killing. And not simply killing the enemy, we had one just standing in line for chow which was "1, 2, 3, attack the chow hall (repeat), Kill the women, Kill the Children, Kill, Kill, Kill 'em All." Constantly using the term "kill" as though it meant nothing was used to desensitize the recruits to the notion of killing and its implications.

As part of instilling what one marine from the same time calls "not just the ability to kill but the lust to kill," boot camp features two weeks of dedicated marksmanship training culminating in the qualification test with the M-16 service rifle.[10] Live-fire exercises make up the second week, when recruits learn to shoot from any position and are expected to hit a human-silhouette still target at five hundred yards. After graduating from recruit training, those marines designated for the infantry go on to the fifty-nine-day Infantry Training Battalion, where they receive far more extensive marksmanship training with both still and moving targets.[11] "Every marine a rifleman," a leading motto of the Marine Corps, betrays its reliance on the basic-level accurate-shooting grunt for its effectiveness as a military force.

Once assigned to an infantry battalion, a rifleman's preparation for deployment to Iraq involved increasingly authentic tactical training—simulation exercises meant to have the look and feel of combat. For Iraq, this simulation often focused on the particular demands of urban warfare. The Marines ran the infantry battalions through Old MOUT [military operation in urban terrain] Town in Camp Pendleton, California—a mockup cityscape with streets of dilapidated blockhouses to simulate the structural challenges of close-quarter fighting. The Marine Air-Ground Center at Twentynine Palms, California, claimed greater authenticity with its two-acre Middle East–like "Combat Town" and live fire exercises and native role players. As the war in Iraq evolved into an insurgency, some infantry battalions rounded out their predeployment training with Security and Stabilization Operations in the former George Air Force Base in Victorville, California, which concentrated on operations amid civilian populations.[12]

In examining the Marines' preparation for combat in Iraq, a method of rehearsal and visualization emerges that gives fuller dimension to the optics of combat. Once in Iraq, combat units found makeshift ways to run through and "see" what they expected to do in the fighting ahead. In preparing for the second battle of Fallujah, for instance, an infantry squad from First Battalion, Eighth Regiment Marines set down "tape houses" in camp—simple outlines in the sand of houses for its fire teams to practice breaching and clearing a building. Another squad built a "schoolhouse" out of scrap wood and sandbags for the same type of last-minute rehearsal before attacking Fallujah. At a higher level, Major General James Mattis, commanding general of the First Marine Division, often spoke of his effort to "visualize the battlefield." In preparation for the March 2003 invasion of Iraq, Mattis and his officers tried to picture the operation in its entirety by arranging its particular parts. With some six thousand multicolored Legos, they represented every unit and vehicle in the division by setting out all the blocks in a Camp Pendleton parking lot.[13]

From high-ranking officers seeing the big picture of an operation to a rifleman visualizing his part in breaching a building, the method of imaging for battle runs deep in Marine Corps culture while drawing from techniques athletes have developed to enhance their performance in competition. During the Iraq invasion, retired major general Ray L. Smith, accompanying the First Marine Division fighting its way north, told an infantry squad to envision the best way to win a fight and then act in concert with that picture before the enemy could do the same. "In all his firefights," according to Marine historian Bing West, "he had developed the habit of mentally running a video through his head, picturing what he expected to see. There was the least confusion when every rifleman had imagined—had thought through—exactly what he was going to do."[14] In another very telling example, Lieutenant Colonel B. P. McCoy, commanding officer of the Third Battalion, Fourth Regiment during Operation Iraqi Freedom, described his marines' "final preparation" for fighting. It "dealt with the act of killing," he later wrote, "the essence of war. Our challenge was to prepare our young warriors to recognize and kill legitimate human targets without hesitation, to understand when the killing had to stop, and to not bear the guilt of killing once the fighting ended. Much like professional athletes mentally imaging themselves performing in competition," McCoy explained, "we actively imaged our Marines . . . through the act of killing." McCoy emphasized the undertaking's practical nature: "This was not an exercise in delusions of grandeur; in fact, it was just the opposite. Imaging must be closely coupled with habits formed in behavioral training"—by which he meant boot camp, infantry battalion training, and urban combat simulations.[15]

Marine imaging in Iraq also meant using high-tech systems of seeing and representing areas of operation, allowing commanders and lower-level officers to overview enemy troop placements, roads and terrain, and other key elements within a battle space. Satellite mapping combined with global positioning systems (GPSs) provided a wide array of intelligence, surveillance, target acquisition, and reconnaissance capacities. New FalconView three-dimensional terrain visualization software provided battalion and company commanders with the ability to plan missions and present video fly-throughs of objective areas from their laptop computers. The system had a significant tactical impact. In the first days of the invasion, First Battalion, Seventh Regiment used FalconView in preparing to take a Rumaylah pumping station in Iraq's southern oil fields. Its 3D digital imagery "enabled the Marines to simulate walking through the pumping station," as Bing West described it. "The Marines enjoyed using the program in their squad bay, and eventually they believed they could have moved through the pumping station blindfolded," he continued: "The National Imagery and Mapping Agency added a three-year-old detailed satellite map product and the Marines pored over every inch of the glossy, visualizing their approach, their dismount from the amtracs, the cuts in the fence, and the attack across the objective."[16] On a grander scale, the Marines utilized an early version of the forebodingly named sensor, the Gorgon Stare, a nine-camera surveillance system that captures motion imagery within a space of two-hundred square miles. As discussed in chapter 1, with the network-centric sensor systems used in the Iraq War, the U.S. military strove for omniscient comprehensive seeing for its air and ground forces.

First and foremost a work on the optics of combat, this study critically describes how the United States projected force in Iraq by fashioning and deploying ways of seeing conducive to the commission of violence in battle. A leading term in military doctrine and political science for expeditionary warfare, *force projection* is adapted here to examine the visual culture of battle and, specifically, to analyze two optical dynamics of American war making in the early twenty-first century. First, military power and killing force flowed from technologies of seeing: the capacity to shine light on an object and, with that, to destroy it. "Light it up": this common Marine command in Iraq enunciates the coterminous nature of perception and destruction.[17] From night vision goggles to the thermal sight of an Abrams tank, from the FalconView tactical imaging system to the cameras of a predator drone: vision-based information technology—extensions of the human eye—determined the Marine Corps' killing power and the American war effort overall.

Second, projection of force goes to the outlook of the marines themselves and what I describe as a culturally constructed eye for battle shaped by the institutional imperatives of the Corps—its history, traditions, myths, training, doctrine, language, and organization—and sharpened by a panoply of early twenty-first-century war pornographies: from hypermasculinist cable TV programming to first-person shooter combat videogames and to YouTube videos aggrandizing face-to-face killing in battle. Chapter 2 focuses on war pornography by closely examining the content of leading YouTube videos made by American marines and also by taking seriously the concept of pornography's causal equation between text, mindset, and behavior. Popular visual culture celebrated combat in a vernacular, masculinist vein while also supporting U.S.

foreign war making by modeling a mindset and outlook of a warrior class still made up of young American men. This book argues that in wartime this particular way of seeing is itself a projection of force, shining a type of light through marines on the ground in Iraq. Its ultimate projection has been physical force and violence against the objects of that sight. My study advances, then, a quite literal meaning of force projection. American military power is predicated on the imaging of violence: the transformation of seeing as primary mode of human perception to seeing as military targeting; and the acting out of killing and destroying, representations of which have already played before the infantry's eyes. While drawing the equation in war between seeing and destroying, the phrase *Light it up* points to the pleasure that marines have found and cultivated in battle.

The "Marine eye for battle" is a historical enterprise, the result of the complex interplay between biological and cultural evolution. The eye can be understood as a type of Cartesian tube, joining the mental world to the physical world, and as the key entity in the tightening loop between the conduct of battle on foreign ground and the cultural privileging of the horror, excitement, and pleasure of fighting and killing for the United States. In considering this loop between culture and war, I argue that early twenty-first-century U.S. war making has realized a near convergence between the two—that is, a near convergence between cultural text and behavior in war, as edited footage of firefights taken by marines with digital cameras is then viewed by nonparticipants on their desktops and by recruits in boot camp. Some of those viewers become new combatants who—equipped with helmet cameras or, say, in the next war, Google Glass (fig. I.1)—will then create a new set of textual referents of combat, fed almost immediately to constantly hungry viewers and "experiencers" of war, thus completing the loop and beginning the cycle again. Historians of military culture need to attend to how the eye feeds and informs "what it's like to be in battle," particularly while studying a period and a people so immersed in a constant cascade of war-based images.[18]

As a historical creation, the Marine eye for battle owes its deadliness in part to the Corps' living legacy of toughness, readiness, and eagerness to fight and kill. Its *First to Fight* slogan—coined by Marine recruiters during the First World War—still speaks to the Corps' interservice identity as a small and selective force always ready for immediate deployment and combat (fig. I.2). *First to Fight* also goes to the long-standing motivation to join the Marines. In addition to wanting to serve, it's an impulse to step outside of the very society (if

Force Projection and the Marine Eye for Battle   9

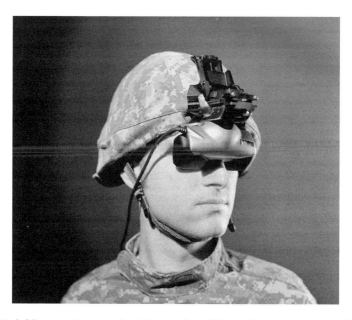

Figure I.1. Soldier wearing virtual reality goggles at Flatworld, a combat simulation system at the University of Southern California's Institute for Creative Technologies. The U.S. military's reliance on high-tech optics has brought a posthuman, cyborgian grain to the infantry, a development that has diminished the hypermasculinity so central to American martial culture. (Lisa Barnard)

only for a time) one is protecting.[19] "They didn't want anything to do with the soft, nasty civilian life," as David J. Danelo describes in *Blood Stripes: The Grunt's View of the War in Iraq* (2003): "They wanted challenge. They wanted to fight. They wanted to bear the burdens others were too weak to shoulder."[20] More than any other U.S. military service, the Marines have approached battle as a test of individual fortitude as well as institutional mettle and reputation. As seen in chapter 3's discussion of the second battle of Fallujah, the Marines prepared for that difficult and brutal fight with no small sense of combat as competition and even play. Above all, perhaps, the Marine eye for battle's effectiveness comes from Corps training and culture having instilled a "killing switch" in its riflemen. "I circled the front sight tip of my rifle around to zero in on him," Corporal Marco Martinez recounts targeting an Iraqi fighter in Baghdad in spring 2003: "With my left eye closed, I got my front sight tip centered in the rear sight and focused on his chest. I took a deep breath and pulled the trigger. The round left the barrel and hit him in the chest, knocking

10 *Light It Up*

Figure I.2. A 1917 United States Marine Corps recruitment poster for the First World War. Always ready for a fight, the Marine Corps solidified its identity as hard-charging infantry during the World War I battle of Belleau Wood. The poster's headline "Tell That to the Marines!" would become a Corps catchphrase, a prideful warning not to cross the United States and its interests. The transformation from adult civilian to marine for world war stands in contrast to early twenty-first-century Marine recruitment posters and commercials that target younger unmolded Americans to join a warrior elite. (United States Marine Corps)

him down . . . 'Got you, motherfucker!' I said." Martinez remembers feeling "no remorse or sympathy" for whom he had just killed: "I had finally become a true Marine."[21]

The problem would come in the long counterinsurgency stage of the Iraq War, when the Marines ordered the riflemen to "turn off the killing switch." Chapter 4 covers this most complex dimension of warfare, when the Marines needed to see Iraqis as human beings—commanding officers explicitly spoke of empathizing with the civilians—while remaining vigilant, aggressive, and violent against fighters living among the noncombatants. Such cognitive, emotional, and perceptual nimbleness is not, by its own admission, the Marines Corps' strong suit, even as its legacy also includes numerous long and difficult interventions as an occupying force and a book, *The Small Wars Manual* (1940), compiling doctrine and lessons learned from fighting counterinsurgencies. A different kind of difficulty has arisen since America's direct military involvement in Iraq ended in 2011 when the Pentagon and the Marines started to consider what the Corps' role would become in the U.S. military force during this century. Greatly compounding that question, as examined in chapter 5, has been the advent of unpiloted aerial vehicles (including predator drones), ground robots for reconnaissance and combat, and the specter of a near military future dominated by posthuman warfighting. Will the human eye be replicated in automated combat systems? Distancing the killer from the killed with technologies of seeing has already brought far from precise firepower, massive civilian casualties, and a certain low-risk ease of making war. Disconnecting killing from human will and volition may bring something altogether different.

CHAPTER ONE

Shock and Awe and Air Power

March 25, 2003. Day 7, Operation Iraqi Freedom.
U.S. Marine Sergeant Colin Keefe couldn't see. Not the horizon. Not his fellow marines of the 3/5. Not even his hand in front of his face. And certainly not Saddam's Fedayeen fighters, who, as Keefe knew, were still "out there." It was "the mother of all sandstorms," as some marines called it, "the Orange Crush" that pounded relentlessly at their bodies, faces, and eyes. Keefe remembers the sensation of sightlessness: "It was pitch black in the middle of the day. . . . You just couldn't see anymore . . . even with your night vision goggles on."

Blindness brought the Marines' invasion of Iraq to a halt. With its objective the seizure of Baghdad—the Corps' longest march ever from the sea—the I Marine Expeditionary Force of some sixty thousand troops had set out north from Kuwait, alongside the Army's V Corps, meeting limited resistance in the southern oil fields and, then, after crossing the Euphrates River, winning a costly battle at Nasiriyah. The next day brought progress north with considerable fighting. Keefe's unit was ambushed by "Fedayeen irregulars," approximately two hundred black-clad militia firing mortars and rocket-propelled grenades from canals south of Diwaniyah. Some marines from the Third Battalion, Fifth Regiment "dismounted" and drove out the Iraqi fighters from the canal trenches on foot with small-arms fire. Keefe remained mobile, shooting and killing Fedayeen with the MK-19 rapid-fire grenade launcher mounted on the back of his Humvee. After one hour of combat and "cleanup," with all the Fedayeen either dead or having "run away," the marines continued on the highway toward Baghdad.

And then the sand started to fly. "We had moved five miles north from the fighting," as Keefe recalls, still taking sporadic fire, when "the sand storm hit and . . . you just couldn't see anymore." Keefe's battalion came to a complete halt. "We parked," Keefe describes: "We got in a three-sixty . . . and then we were just sitting there." With the distinct possibility of the Fedayeen fighters attacking again, Keefe remembers the storm—some six hours of being stuck and blind—as his most "frightening" time in Iraq.[1]

This book examines what happened when Keefe and the United States Marines could see again. How did the Marines utilize light, optics, and perspective in invading, seizing, and occupying Iraq from 2003 to 2011? What was the relationship in the Iraq War between the U.S. Marines' capacity to see and to kill in combat? Exactly what happened when the lights went back on for one of the most powerful invasion forces in history? In the first instance, and throughout the war, the answers to these questions turn on air power and its ever-expanding techniques and systems of seeing and targeting the enemy from high above and across great distances.

The air power projected on Iraq in spring 2003 was the culmination of what historian Michael Sherry has described as the long-standing embrace by the U.S. military of "bombing as an abstract idea." Starting in World War II, American policy makers and military officials have exhibited, in Sherry's well-known term, a "technological fanaticism" over the methods of destroying people and other targets from increasingly more advanced and removed weapon systems; a "casual brutality" has accompanied the evolution of American air power—less an active bloodlust for massive casualties, as seen in British and German "morale bombing" during World War II, and more a willful blindness or indifference to the human cost of its doctrine and expertise.[2] The vast limitations of this superior military capacity came into clear view in the Iraq War, as force projection from the air brought high civilian casualties and, with that, the emergence of new enemies. As Commanding General Tommy Franks understood the equation between air power and victory in the spring 2003 invasion of Iraq, "The operation is longer the less kinetic we put into Baghdad; the more kinetic, the strategic exposure goes up."[3] With an entire ground invasion force of 195,000 troops, including British troops, the United States banked on a new version of the old promise that technology could replace boots on the ground and thereby reduce risk to both American troops and public support for the military venture itself.

Network-Centric Warfare, Sensors, and Total Situational Awareness

The Marines, along with the other U.S. armed forces, started fighting in Iraq amid a broadly gauged *transformation* toward *network-centric warfare*, a veritable *revolution in military affairs* that marked, for its proponents, a *new way of war*.[4]

Military theorists' penchant for grandiloquent terms was never more evident than in the years following the 1991 Persian Gulf War, when advances in digital communications, satellite global positioning, laser-guided munitions, and a host of related technologies combined to promise worldwide force projection and dominance of American enemies via remote sensors, instantaneous communication, and airborne shooters. Large land armies, under this view, were no longer necessary. Secretary of Defense Donald Rumsfeld sought to implement such *immaculate warfare* in the post–September 11 invasions of Afghanistan and Iraq, the latter operation seeing a 1:6 force ratio of U.S. and Coalition troops to their counterparts.[5] The March 2003 Iraqi invasion met with great tactical success, as U.S. Marine, U.S. Army, and British forces rapidly won military control of Iraq and its capital by April 6, 2003. But the high-tech-driven speed and precision with which Operation Iraqi Freedom took hold of the country could do little to thwart the insurgency that started in 2003 and that remains today. In strategic terms, *shock and awe* was an abject failure.

To examine network-centric warfare—its definition and basic implementation during the Iraq invasion—is to appreciate how the concept rests on superior sight and perception. In the imaginations of military policy makers and defense contractors, network-centric warfare is meant to achieve a god's-eye view of the battlefield and, indeed, of the planet, mastering terrestrial information from air and space, knowing "all things at all times everywhere in the world" via a "staring, all-seeing eye."[6] In 2004 the Department of Defense officially defined network-centric warfare in more sober terms: "An information superiority-enabled concept of operations that generates increased combat power by networking sensors, decision makers, and shooters to achieve shared awareness, increased speed of command, higher tempo of operations, greater lethality, increased survivability, and a degree of self-synchronization." Even in this Defense Department document, though, one detects the giant-eyeball-in-space ideal that drives network-centric warfare doctrine and theory. "In essence," the 2004 definition concludes, "network-centric warfare translates information superiority into combat power by effectively linking knowledgeable entities into battlespace."[7] Because the "knowledgeable entity" category fits pretty much every person, place, and material thing, it can be said that network-centric warfare holds out nothing less than omniscience as a realizable goal for the U.S. military.

Network-centric warfare can be understood as a "system of systems" that begins with comprehensive perception and ends with precision destructive power; in the middle stand the decision makers, the battlefield commanders, whose "enhanced situational awareness" assures tactical advantage over the enemy. To emphasize, within network-centric warfare doctrine and its constituent parts, military dominance manifests through superior perception and cognition. Firepower and troop strength are seldom addressed. Rather, network-centric warfare seeks to limit or, as in shock and awe, "overload an adversary's perceptions," while achieving omniscience for its decision makers.[8] Network-centric warfare "ensures," an an article in *Marine Corps Gazette* explains, "that the MAGTF [Marine Air Ground Task Force] commander has total visibility of the battlespace."[9] Network-centric warfare comprises an information circuit that cuts through the fog of war. Full perception and total situational awareness mean increased speed of command, which, in turn, brings greater fog of war—that is, significantly decreased perception—to the enemy.

The sensors employed during the spring 2003 Iraq invasion ranged from satellite systems in space to hand-launched unmanned aerial vehicles flying just a few hundred feet over enemy lines. From outer space, imaging radar satellites located and followed Iraqi Republican Guard forces so they could be targeted and destroyed by air power. Satellites intercepted Iraqi command communications and photographed enemy troop placements.[10] The satellite-based global positioning system (GPS) was invaluable to U.S. operations in the Iraq War, marking a significant difference from sensory capability during the Gulf War. In preparation for Operation Iraqi Freedom, the Pentagon enhanced the GPS satellite navigation system to provide accuracy from what had been sixteen meters to within three meters for precision-guided munitions, aircraft, and ground forces. GPS satellites also enabled the new Blue Force Tracking system. Used extensively by U.S. Marine and Army ground units during the invasion, the Blue Force tracker identified on laptop computer screens the location of friendly troops relative to the viewer's position and reduced friendly fire in Iraq, the leading cause of American casualties during the Gulf War.[11]

Operation Iraqi Freedom also saw the introduction of the Global Hawk unmanned aerial vehicle (UAV), a reconnaissance drone with two real-time "sensor packages"—synthetic aperture radar and the electro-optical-infrared system. The Joint Surveillance Target Attack Radar System (STARS), first used in the Gulf War, had demonstrated the value of real-time images of battle space. The Pentagon and commanders weren't satisfied, though. They wanted twenty-four-hour coverage of Iraqi forces. The Global Hawk UAV was developed, as the Defense Science Board put it in 1993, to "fix the problems exposed in Desert Storm." After flying over Afghanistan in Operation Enduring Freedom, the Global Hawk AV-3 arrived over Iraq in early March 2003, when it contributed to the prewar campaign to destroy Iraq's air defenses. The Global Hawk then fixed on Republican Guard locations, as its synthetic aperture radar generated detailed "unblinking" coverage from as far as one hundred miles. In Operation Iraqi Freedom, AV-3 typically set up over "kill boxes" some three hours before an attack, relaying to fighter pilots and ground commanders images of Iraqi vehicles and troop placements whose resolution was high enough to spot enemy fighters on foot. Global Hawk also carried one of the few sensor systems that could see through the late March "orange crush" sandstorm. While the sand and dust rendered AV-3's infrared sensors useless, its radar could still track the Republican Guard below.[12]

Among the sensor systems employed in the Iraq invasion, Joint STARS was central to the effort by tracking ground vehicles and troops on foot and relaying tactical images to battle commanders, information that helped coordinate U.S. forces in the operation. In the Iraq War, Joint STARS was housed in E-8C aircraft (modified Boeing 707s), with a crew of four airmen and radar that could detect stationary and moving objects. Over and above the Global Hawk UAV, Joint STARS's communication system relayed data directly to ground stations and battle commanders. It accepted radio and text-messaging data requests from ground units. Air National Guard major Thomas Grabowski, senior director on the E-8Cs, described the Joint STARS as the "911 call" for ground troops: "If one of these convoys gets in trouble—they break down, they have troops in contact, small-arms fire or any type of a problem—they call us," Major Grabowski said. "We're like the 'On-Star' for the ground commander."[13]

For the marines on the ground in Iraq, Joint STARS, along with the other airborne sensors and the digital communications network that accompanied them, meant that, as the *Marine Corps Gazette* put it in 2002, "the opportunity to exploit this 'digital dimension' of the battlefield, once enjoyed mainly by those in senior headquarters, now extends down to the last tactical mile." Regiment and battalion commanders were now linked in. "Beyond line of sight" of enemy forces, the images and other data accessed by Marine commanders translated into "increased SA [situational awareness] at the company and platoon level. This increases a commander's ability to rapidly receive information from sensors and systems and deliver it to weapons platforms," as the *Marine Corps Gazette* concluded, adopting one of the Marines' favorite terms, "by using the network as a 'force multiplier.'" In more local fashion, platoon and squad leaders depended on images from the Dragon Eye UAV, an "over-the-next-hill" tactical reconnaissance and surveillance platform that provided both real-time moving pictures and "click-capture" stills of enemy positions up to five kilometers from its hand and bungee-cord launch.[14]

Achieving Rapid Dominance in Iraq

The March 2003 opening assault on Baghdad and Saddam's forces was organized and carried out with the goal of "achieving rapid dominance" through "shock and awe." More than a caption CNN used to promote its live attack-on-Baghdad coverage, shock and awe connected back to military doctrine formulated in a 1996 National Defense University study group. Authored by Harlan K. Ullman and James P. Wade, the study group's summary statement (later published in book form) situated its proposals within the revolution in military affairs: "What is radically different in Rapid Dominance is the comprehensive system assemblage and integration of many evolving and even revolutionary technical advances in dominant battlefield awareness squared." Those advances include "materials application, sensor and signature control, computer and bioengineering applied to massive amounts of data, that enable weapon application with simultaneity, precision, and lethality that to date have not been applied as a total system." For Ullman and Wade, shock and awe would exceed the preexistent "decisive force" standard by an order of magnitude. Within the doctrine's terminology, "dominance" denotes "the ability to control a situation totally," while "rapid" means nothing less than "'own[ing]' the dimension of time—moving more quickly than an opponent, operating within his decision cycle, and resolving conflict favorably in a short period of time." The intent is force projection that, in effect, supersedes the enemy's military defenses by attacking the adversary's "will, perception, and understanding."[15] With CNN viewers in awe of the network's shots of cataclysmic explosions and fireballs curling up into the dark evening sky, U.S. forces attempted to bomb Iraq into shock, a state of dumbfounded paralysis that would allow the country to be vanquished before mustering its defenses.

In their study "Shock and Awe: Achieving Rapid Dominance," Ullman and Wade discuss eliminating an adversary's perception in something close to literal fashion. They're presumably speaking figuratively while describing how shock and awe involves "get[ting] into the minds of the adversary far more deeply than we have in the past." But shock and awe also "means literally 'turning on and off' the 'lights' that enable a potential aggressor to see or appreciate the conditions and events concerning his forces and, ultimately, his society." In struggling to find a practical example of a weapon system of such magnitude that would turn off the enemy's lights, Ullman and Wade mention the atomic bombs dropped on Hiroshima and Nagasaki: "The Japanese simply could not comprehend the destructive power carried by a single airplane. This incomprehension produced a state of awe." They also imagine a huge battlefield stun grenade causing the desired effect of sightless shock and awe: "The stun grenade produces blinding light and deafening noise. The result shocks and confuses the adversary and makes him senseless." Its aim "is to produce so much light and sound or the converse, to deprive the adversary of all senses, and therefore to disable and to disarm. Without senses, the adversary becomes impotent and entirely vulnerable." Ullman and Wade speak positively of a "high altitude nuclear detonation" that would knock out the enemy's electronic systems and its capacity to see. Shock and awe doctrine seeks total control. While network-centric warfare calls for omniscience through its sensor systems, shock and awe strives for omnipotence: "With our technology and people force, why should the United States . . . ever put ourselves into a situation where we have one soldier against another."[16]

Even though the United States utilized neither giant stun grenades nor high-altitude nuclear warheads in its Iraq invasion, the operation, in the word of James Wade, "really" did follow "Shock and Awe: Achieving Rapid Dominance."[17] Wade and Ullman's text promotes "stealthy bombers," "strike aircraft," and "cruise missiles" as the most practical weapons systems for a shock and awe operation.[18] And, indeed, with a total ground force of only 195,000 troops, these systems led the way in the spring 2003 Operation Iraqi Freedom. Although the air campaign against Saddam Hussein actually began several months earlier, during the summer of 2002, when UN-sanctioned air patrols started to attack Iraq's radar and communications systems and surface to air missiles and to disrupt its air defense system. In July 2002 U.S. air war commander T. Michael Moseley authorized pinpointed strikes against targets in southern Iraq. Dubbed Southern Focus, the secret plan accounted for more than 600 bombs dropped on 391 targets, including a fiber optics communications network running between Baghdad, Basra, and Nasiriyah.[19]

Under General Tommy Franks's initial plan, Operation Iraqi Freedom would begin with five days of devastating air strikes on some 437 targets. In early March 2003, just weeks before the invasion, however, General Franks and President George W. Bush's administration dropped the extended air campaign before sending in ground troops. Franks and the Pentagon predicted that Saddam would destroy his own oil fields in southern Iraq as soon as the air strikes began. Since that oil was meant to finance both the war and the Iraqi recovery, U.S. command decided that ground troops would have to seize the southern fields in the opening phase of the war.[20] In another strategic matter, the Bush administration worried that the shock and awe air strikes would be so devastating for Iraqi civilians the nation would turn against its would-be liberators. Ullman and Wade touch on the concern in their 1996 publication on shock and awe, asking dully, "Would the concept of Rapid Dominance offend and generate counterproductive public relations backlash from those who believe force should only be used as a last resort and then with a measurable degree of proportionality?"[21] Operation Iraqi Freedom's originally planned air campaign had called for a high degree of disproportionality. It included twenty-two "high-CD" targets, the Pentagon's term for the possibility of more than thirty civilian casualties.[22] Implementing a new software program tactlessly nicknamed "bugsplat" that predicted destruction from air strikes, the Pentagon tried to split the difference between military victory through shock and awe and the likelihood of killing, terrorizing, and making enemies of the Iraqi people.

To what extent the United States limited the destruction and civilian casualties of shock and awe in the March 2003 attack on Iraq is very much a matter of perspective. The proposition of muting a military operation doctrinally based on overwhelming and disorientating force would appear to be a contradiction in terms. Yet, with the air strikes cut from five days to one day, and with U.S. strategic interest in limiting civilian casualties, it's incontrovertible that, well, Operation Iraqi Freedom could have been worse. Secretary of State Donald Rumsfeld bristled over the press's comparison of the air campaign to the leveling of Dresden and other German cities during World War II. "The targeting capabilities and the care that goes into targeting to see that the precise targets are struck and that other targets are not struck is as impressive as anything anyone could see," Rumsfeld claimed in his first press briefing after the operation had started. "The humanity that goes into it," Rumsfeld continued, "to see . . . that it's done in a way, and in a manner, and in a direction and with a weapon that is appropriate to that very particularized target . . . I think the comparison is unfortunate and inaccurate."[23] Early twenty-first-century precision-guided munitions did make major improvements in accuracy over World War II carpet bombing and, for that matter, over the Gulf War's so-called smart bombs. But civilian casualties did occur. The precise number of dead and wounded Iraqi civilians from the invasion is hard to come by, but the Iraq Body Count study, typically conservative in its estimates, lists 4,613 Iraqi civilians killed by violence during the first two weeks of Operation Iraqi Freedom. The 2004 *Lancet* study, named after the highly reputable British medical journal in which it was published, estimated that, from the beginning of the invasion to September 16, 2004, more than 98,000 excess deaths (over the number that would have happened without war) occurred in Iraq.[24]

Despite the decisions made in the weeks leading up to the operation to try to limit civilian casualties, U.S. command planned the Iraq invasion within a framework of strategic air power, a historically imprecise and exceptionally destructive approach to war making. Shock and awe, critics charged, is Guernica under a new name. "Shock and awe is what air forces have been doing since World War I," as political scientist Robert Pape put it: "That is always the plan."[25] Shock and awe, from the perspective of Iraqis on the ground, must have been terrorizing as well as deadly.

"Shock will be introduced at 2100" is how General Franks announced to his commanders the 9:00 p.m., March 21, 2003, beginning of the invasion. Strikes from the air would come in three nearly simultaneous waves and from a wide array of shooters. First, the U.S. Navy launched more than eight hundred Tomahawk cruise missiles, each one carrying 1,000-pound bombs. Next came the U.S. Air Force's F-117 stealth attack aircraft, F-15C and F-16 fighters, and Navy and Marine F/A-18 fighters—all dropping precision-guided bombs. And, finally, B-2 stealth bombers dropped JDAMs (joint direct air munitions). In all, the one-day air campaign against Iraq included more than seventeen hundred sorties.[26]

Kill Boxes, LITENING Pods, and the Third Marine Aircraft Wing

For all its sensors, shooters, and bombs, the shock and awe campaign did not produce the immediate collapse of Iraq's military. For the ground war that still had to be fought and won, U.S. air power turned devastatingly tactical, attacking Iraqi armored and infantry divisions with, among other things, 500- and 1,000-pound laser-guided bombs and the Air Force's A-10 Warthog gunship with its 30 mm rotary cannon. For the I Marine Expeditionary Force (I MEF), tactical air power meant the Third Aircraft Wing: with 435 aircraft and 15,000 marines, it made up the largest standing Marine aviation element to go to war since Vietnam. The Third Aircraft Wing's operations in this "Opening Gambit" of the Iraq War mark a high point in Marine aviation history. Crossing into Iraq, Major General James Mattis of the First Marine Division decided to rely more on air power than artillery to weaken Saddam's forces. Marine F/A-18 Hornets and AV-8 Harrier jets worked in "kill boxes," three-dimensional free-fire zones that organized interdiction and deep battle targets ahead of friendly lines. Technological advancements in onboard optics allowed pilots to maintain high altitudes while employing lasers and other precision guidance munitions systems to deliver ordnance with unprecedented accuracy. Avoiding unnecessary collateral damage remained a strategic concern, although how well the Marines realized that goal remains a matter of perspective. In both the initial attack and the occupation that followed it, Marine aviation leveled extraordinary firepower against Iraq and its people. "We had this outfit called the Third Marine Air Wing," Lieutenant General James Conway, commanding general of I MEF, said in a 2004 interview broadcast on the History Channel, "and I will tell you, frankly, it's a killing machine."[27]

The modern U.S. Marines and Marine aviation started out together in the early twentieth-century so-called Banana Wars in Central America and then came of age together in the Pacific Theater during World War II, when the Corps' exceptional capacity for expeditionary warfare coalesced around amphibious landing and close air support. Operating from aircraft carriers as well as land bases, Marine fighters would fly in coordination with attacking infantry, focusing fire on targets "close" to friendly ground troops with greater precision than naval guns or artillery could manage. By the Vietnam War, the Marines had integrated rotary-wing aircraft into troop transport and close air support and, had coined MAGTF (pronounced *mag*-TAFF) to denote the Corps' most basic combat unit: the Marine Air Ground Task Force. "MAGTF" became central to Corps culture, doctrine, and discourse as the acronym marked the elemental nature of air-ground teams in Marine operations. Oral histories from the Iraq War toss off the term as a synonym for Marine warfighting. For example, in April 2003 Lieutenant Colonel Willard Buhle, commander of the Third Battalion, First Regiment, described his unit's efforts during the Iraq invasion as "focused on bringing our combat power to bear, fighting the MAGTF, and we're good at it."[28] During the Opening Gambit the Third Aircraft Wing's fifty-eight attack Cobra helicopters provided forward armed reconnaissance of advance-route roads as well as cannon and rocket close air support. Overall, the Marines marked the air element in MAGTF an unparalleled success in Operation Iraqi Freedom. "We flew more than 10,000 sorties and dropped more than 6.5 million pounds of ordnance," said Major General James Amos, commanding general of the Third Aircraft Wing: "In the 28.5 days of missions, we destroyed eight Iraqi divisions, two of which were the elite Republican Guards."[29]

In carrying out their varied Operation Iraqi Freedom missions, Third Aircraft Wing pilots relied on an array of sensors: from commonly referred to "eyeballs" to the Advanced Tactical Aerial Reconnaissance System (ATARS), and from forward air controllers attached to Marine ground forces to the LITENING Pod. Many Marine pilots who had flown over Iraq during the prewar Operation Southern Watch commented on how those earlier missions added to their effectiveness during the March 2003 attack. "In perspective," Major J. Scott Wedemeyer stated, "it was good to fly over enemy territory and get a God's eye view of the place."[30] Wedemeyer and other F/A-18 Hornet pilots spoke about identifying targets in kill boxes with their bare eye.[31] In interviews taken after Operation Iraqi Freedom in the United States Marine Corps History Division Oral History Project, pilots exclaimed over the effectiveness of new optical and sensor systems. Set in the nose of some F/A-18 Hornets and AV-8 Harriers, the Advanced Tactical Aerial Reconnaissance System used infrared and visible light sensors to acquire imagery of target areas, store that imagery, and transmit it to Common Imaging Ground/Surface Stations, where, in turn, it would be distributed to and viewed by pilots preparing for attack missions. Pilot interviews also documented their routine contact with Forward Air Controllers who would target and direct fire from the air, sometimes calling off strikes at the last moment because of potential friendly casualties.[32] According to the postoperation interviews, by far the most important targeting system for Marine attack pilots was the LITENING Pod, what Lieutenant Colonel Clyde Woltman, an AV-8 Harrier pilot, described as a "totally different animal" from what was used during the Gulf War. "The pod is, without a doubt," Woltman added, "the best targeting system in the world."[33]

Mounted externally on the bottom of the aircraft, LITENING is a multisensor laser target-designating and navigation system that enables pilots to detect, track, and identify ground targets for highly accurate delivery of conventional and precision-guided munitions. Its effectiveness as a targeting system during Operation Iraqi Freedom was due to the clarity of vision Marine pilots gained through the Pod's third-generation forward-looking infrared (FLIR) thermal sensor for nighttime missions and its charge-coupled device television (CCD-TV) visual sensor during the daytime. The LITENING Pod's FLIR and

CCD-TV sensors both enjoyed a 750% increase in optical capability over its predecessor.[34] But it was the daytime visuals over which most pilots exclaimed, as the CCD-TV played on a video screen in front of the pilot. "[It's] obviously the best technology we have right now as far as IDing targets," according to AV-8B Harrier pilot Captain Jennifer Dolan; "during the day it is absolutely amazing what you can see from what altitudes and what distances . . . I mean it is almost just like you are flying over the target looking at it with your eyeballs the video is so good." Dolan detailed that "in some instances we have people flying around overhead the target 25,000 feet during the day and you can watch people get in and out of their vehicles."[35] In addition to high-altitude clarity of vision, the LITENING Pod gave Harrier pilots the unique capability to "paint" their own targets with a laser; independent from forward air controllers, they could identify and mark ground targets and drop large ordnance with exceptional accuracy.

With the LITENING Pod, as Captain Dolan put it in April 2003, "the Harrier mission is changing exponentially." One difference was a new Intelligence, Surveillance, and Reconnaissance (ISR) mission for the Harrier. Lieutenant Colonel Kevin S. Vest, commanding officer of Marine Attack Squadron 211, explained that during Operation Iraqi Freedom the LITENING Pod in the Harrier amounted to "a new sensor" for "the MEU" (Marine Expeditionary Unit): "What it gave the MEU commander was the ability to launch two AV-8s off of the same ship . . . and send them perhaps a couple of hundred miles away . . . to go look at a standing at high altitude and not be detected"; with the pod recording what the pilot saw on the CCD-TV video screen, the commander would "then have those aircraft back and actually look at the tape all in a one hour time span." The biggest tactical change the LITENING Pod brought to Harrier pilots was to center their mission well north of advancing Marine ground forces and in kill boxes where they targeted Iraqi artillery, tanks, and troops. "The mission that we ended up doing was quite different than the traditional close air support role" Dolan "thought it was going to be." Colonel Vest reported that "at least the first two weeks of the war" for his squadron "were all AI interdiction-type missions": orders would be "fly to a kill box," and "if you can find any targets, any legitimate targets, you are clear to engage them." After identifying a tank or "artillery piece" with the LITENING Pod, Vest continued, "we put a laser on it and put a GBU-16 (Guided Bomb Unit) . . . into the target. Basically, with the targeting pod and the GBUs, you could hit it and you could kill it with one bomb."[36]

As for close air support, the LITENING Pod made one significant contribution, although perhaps it should fall under ISR (intelligence, surveillance, and reconnaissance). Because of LITENING's enhanced optics and data-link system, Harrier pilots could provide Marine ground forces with detailed information about enemy units' location, size, and disposition. CCD-TV video of what the pilot saw in the cockpit could be sent in "real time" from the targeting pod to the laptop computer of forward air controllers and ground commanders. LITENING's air sensor to ground unit capability was first implemented in Operation Iraqi Freedom by Marine Attack Squadron 214 in support of Task Force Tarawa's battle in and around the city of An Nasiriyah, where a forward air controller, using the video feed from the Harrier, successfully targeted and engaged enemy units with three laser-guided bombs. LITENING's downlink feature would be called a "revolutionary" development in MAGTF tactics.[37] Sharing a wide-frame bird's-eye view of the enemy to ground command reduced the likelihood of harm to friendly forces and Iraqi noncombatants.

The 2003 USMC Oral History Project interviews among Marine pilots flying in Operation Iraqi Freedom divulge considerable institutional effort to minimize risk—to the pilots themselves (through altitude restrictions) and to U.S. ground forces and Iraqi civilians (through new "positive ID" requirements, for instance, and by calling off air strikes on targets too likely to cause unintended casualties). "From a pilot's perspective," as Lieutenant Colonel Vest stated, the operation "was really a departure from some of the ways we are originally trained." It truly had been a matter of changing perspective, Vest emphasized: "I know my eyes and the pilot's eyes, our training is to go out and destroy things as attack pilots." Vest and others felt "that our whole job is to be there to protect Marines" in providing "close air support." As squadron commander, Vest spoke of how "several of my pilots" were "very "frustrate[ed]" in not being able to carry out that role. He described in detail one "incident" in early April 2003 when a Marine ground unit was patrolling Baghdad and started "taking sniper fire from a mosque." With one marine killed and two more injured, and his Harrier squadron overhead, Vest's "pilots could hear the marines calling for help on the ground." And it was the ground unit's officer, as Vest recounted, "who said 'Wait a second, we can't use fixed-wing aviation in here. We have too many friendlies in close and I can't put a 500 pound bomb into the side of the Mosque to kill a sniper, so we are going to have to find another way to do this.'" Vest elaborated on what this incident meant to his pilots. "It was only after returning to the ship where you could really put things in perspective," Vest said: "The Marines on the ground were doing a movement to contact." There was an inherent limit to what the attack pilots could do in that situation:

> Basically they were moving through the streets of Baghdad trying to get fire from pockets of Iraqi resistance that were still located in the city and one Marine did die and ten were injured in this particular incident but had we used an AV-8 and put a GBU 12, a 500-pound laser guided missile into the mosque . . . one most likely would have injured some marines and of course we would have been causing a tremendous amount of collateral damage to an important religious site in Iraq and possibly also killing many civilians. So we would have killed the sniper. We would potentially win the battle but lose the war and that was a hard concept for the pilots to understand at first until you've gone out and experienced it because it's just not what we are used to doing.[38]

Whether Lieutenant Colonel Vest worried about civilian casualties as much as losing Marines or destroying the mosque is open to interpretation. It's nevertheless clear that amid the "killing machine" that was the Marines' Third Aircraft Wing, strategic concerns over collateral damage limited both the aggressiveness structured into attack pilot training and the destructive capacity of the aircrafts' weapon systems and munitions.

To what extent these limitations made a difference is, again, very much a matter of perspective. Iraqi civilian casualties did occur during Operation Iraqi Freedom, many of them from air power. To say that there weren't as many as there could have been is at once true and to some extent beside the point if one focuses on the oversized calibration of U.S. firepower for an invasion of a country whose urban population was putatively meant to go unharmed and be liberated. Bing West recalled how Marine infantry laughed when they heard the word "precision" used to describe cruise missile and 1,000-pound bomb air strikes in the opening weeks of the war. "They knew better," West wrote: "They were the ones who agonized over firing five .50 caliber rounds weighing two pounds, because they saw the damage the rounds inflicted."[39] The indiscriminate nature of American force projection from the air only grew in the months and years ahead. Especially before 2007, which marked an effort to implement new counterinsurgency tactics (as examined in chapter 4), the United States used air power to terrorize Iraqi civilians. That's how many Iraqis saw and experienced it. "There is no limit to American aggression," a sheikh from Baquba recounted in 2005: "The fighter jets fly so low over our city that you can see the pilots sitting in the cockpit. The helicopters fly even lower, so low, and aim their guns at the people and this terrifies everyone. How can humans live like this? Even our animals, the chickens and sheep are frightened by this. We don't know why they do this to us."[40]

In turning back now to what Sergeant Colin Keefe saw and did after the great sandstorm ended in late March 2003, it's crucial to recognize how U.S. air power still helped shape the conditions and conduct of battle on the ground. In early April 2003, after the fall of Baghdad, a Marine task force moved one hundred miles north, to capture Tikrit, Saddam's hometown. The Marines encountered virtually no resistance. In the city streets, veteran Iraqi soldiers in Saddam's Republican Guard could be found in street clothes and fixated on the intensity of American bombing. "We're not cowards," one explained. "But there's no point in fighting when the Americans have this aviation," he continued, "and there is no way we can win. We would just be killed."[41] Tactical air power, in concert with a wide array of warfighting technologies that dramatically increased the optics of Marine infantry, along with the organization, skill, and firepower of those ground forces, all contributed to the great military success of Operation Iraqi Freedom.

"Keep Your Eyes Out," Fair Fighting, and Memories of Killing

Today, when Colin Keefe remembers his actions in the March 2003 invasion, he wants to connect that operation and the war overall to a positive historical direction or storyline for Iraq and its people: "It's important to me that my service there meant something, accomplished something, served the greater good." Speaking in 2014, amid increasing sectarian warfare in Iraq, he doesn't have a lot to work with. And he knows it. After honorable discharge from the Marines in summer 2003, Keefe graduated with honors in History from Lehigh University, completed his J.D. from Georgetown Law School, took a job with a large Wall Street firm, and now practices law in suburban eastern Pennsylvania, where he and his wife are raising three children. Since meeting him in fall 2003, I've always been struck by the combination of thoughtfulness and matter-of-factness in Keefe's schoolwork, memory of the Marines and the Iraq War, and in his person. And yet he's clearly straining in trying to find a "larger good" in the war. Given the long-standing and violent split between Sunni and Shia Muslims, Keefe conjectures, it's "at least possible, probable that Iraq would be tearing itself to pieces as Syria is now. Maybe our invasion and occupation bore some of the brunt of that conflict." Our conversation doesn't pause over the point that "tearing itself to pieces" is exactly what's happening in Iraq; we don't address the question of whether the Iraq War may have actually hastened the civil war in Syria; instead, Keefe concentrates on what's behind his need to find instrumental value in the U.S. war effort. "My thoughts and struggles over the Iraq War revolve around my killing of Iraqis," Keefe explains; clear and continual images of those dead "affect me more deeply than any other memories." He concludes with characteristic straightforwardness: "It would make my service more difficult to live with if I had killed those people for no reason."[42]

Keefe has many thoughts, memories, and feelings about the war. Alongside the "guilt" is considerable pride. He has pride in his enlistment ("it's kind of a tradition in my family to join the military, so I did"); pride in the Marines as a fighting force ("I mean, the Marines are noticeably better at what they do than the Army"); pride in himself, his training, and his capacity in combat ("I was pretty confident of my ability to shoot each man before he had shot me"); pride in the MK-19 automatic grenade launcher he used in the invasion ("a big old gun which sits on top of a Hummer shoot[ing] forty millimeter grenades out to about a range of about a mile"); and pride in his ability to handle that powerful and demanding weapon, which circles back to the longer-lasting guilt over the damage he did with it.[43] His mind's eye helps him recall the excitement of crossing the berm into Iraq to start the invasion. Moving north toward Baghdad, and with his platoon part of a Combined Anti-Armor Attack Team (CAAT), he would largely "ride around in Hummers" and "scout the way and lead the way for the rest of the battalion. So, a lot of times," as Keefe describes the tip of the spear, I was "just kind of going forward into the unknown." He doesn't remember being afraid: "I guess I had a lot of adrenaline going, and a lot of it seemed, as weird as this sounds, almost fun." Like so many marines and soldiers before him, Keefe welcomed the opportunity to see what he could do in combat: "I was just doing my job, keeping my eyes open, looking around, and just waiting to shoot things before they shot me."[44]

Keefe's emphasis on shooting first reflects both Marine Corps infantry doctrine—including "immediate action upon contact with the enemy"—and his investment in the competitive aspect of combat. Compared to sports, "combat is much more mental," Keefe insists. "You've gotta see what the other guy is doing first," he continues; "there's so much more to keep track of, it's much more complex than any sport." Combat does come down, though, to "pitting your skill against another guy." On this point Keefe is resolute. And in such head-to-head competition, Keefe and his fellow marines believe they enjoyed one supreme advantage: "It's just that we shoot so much better. The Marines are so much more accurate than the Iraqis when it comes to firefights that we just always hit them before they hit us." The most remarkable point here, for Keefe, and for so many other marines in the invasion, is just how bad the Iraqis seemed to be. "Gradually," as Bing West documents in his account of the Marine "march up" to Baghdad, "everyone on the front lines seemed to be developing skepticism about, if not disdain for, Iraqi marksmanship."[45] Or, as Keefe describes: "It's almost odd, they just cannot shoot straight. . . . They're horrible, horrible!" More than simple Marine pride or reflexive disdain for the enemy, Keefe's observation is a key part of his memory and original perception of battle. Keefe says he owes his life to poor Iraqi marksmanship. The conviction that "it's more that we're better shots than our weapons are superior" structures his understanding of Marine success in Operation Iraqi Freedom.[46] There *is* more to the story. At the same time, Keefe's perspective can be taken as true and important—that is, as accurate and consistent with his experience and contributive to historical understanding of the Marines' optics of combat in the Iraq War.

Being a bad shot is, in part at least, a relative condition. Iraqi forces (including Syrian and other foreign fighters) may have been, on the whole, worse shots than marines. That the United States Marine Corps has long prided itself on rifle training and proficiency and excellence in shooting certainly informs this issue of Iraqi marksmanship. In subjective terms, the pride and culture behind the Corps' maxim "Every Marine a Rifleman" could influence the denigration of an enemy's marksmanship. More materially, though, Marine infantry marksmanship stems from hundreds of hours of firing-range training—from boot camp, to Infantry Training Battalion, and to live-fire qualification skills training within a Weapons company. Even the elite Iraqi Republican Guard troops had nowhere near the same level of training—to say nothing of the regular Iraqi Army, Fedayeen Saddam, and irregular militia. As discussed in this book's introduction, since the mid-twentieth century, the U.S. Marines and U.S. Army have used, among other conditioning techniques, human-shaped targets for weapons training in an effort to avoid the high nonfiring rate reported by S. L. A. Marshall in World War II. Did Iraqi infantry training devise equally effective infantry training methods for overcoming what David Grossman has posited as the instinct against killing another human being? If not, one reading of the many marine accounts of the Iraqis' astonishingly bad shooting is that they aimed high or otherwise *tried* to miss.

There are less speculative explanations for what may have been a significant difference in accuracy between U.S. Marine and Iraqi shooters. Given the firepower of the initial American shock and awe attack from the air, those surviving Iraqi infantry who remained in rank and opposed the invading ground forces may very well have been less than steady and focused on the task at hand. The concussive and disorienting effect of 1,000-pound bombs from U.S. warplanes should not be dismissed; nor should the impact from the ongoing close air support of the Marines' march north. Those Iraqi troops who did not desert may have been forced in one way or another to stay; to take aim, shoot, and kill one American among such an overpowering enemy may have made little sense; better to sink into the civilian population and live to fight another day and in another way. In a more material vein, to understand the relative inaccuracy of Iraqi shooting (or the possibility thereof), we need to return to the high technology that drove Operation Iraqi Freedom and further appreciate its capacity to see and target across great distances. The bulwark of U.S. sensors, trackers, and sights—from satellite mapping of battlespace to night vision goggles—greatly influenced the ground war and the Marine grunt's capacity for killing. While collapsing space and time for American forces, high-tech optics of combat—per shock and awe theory—deteriorated, in reciprocal fashion, the enemy's perception and understanding. Sergeant Keefe had little occasion or reason to consider this overarching advantage in force projection; the complexity, challenge, and deadliness of fighting through Fedayeen ambushes demanded his focus. But those opposing Fedayeen fighters and other Iraqi troops were to some extent aware of and, most importantly, subject to combined American superiority in firepower and optics. It reduced the Iraqis' will to fight as well as their ability to shoot straight.

One of the farthest-reaching weapons platforms in the Operation Iraqi Freedom ground force was the M1 Abrams tank. With devastating firepower from its rifled cannon and three machine guns, the Abrams targeted objects through thermal and day sights at eight thousand meters. Its accuracy at great distance was ensured by a digital fire control computer. The tank gunner only needed to set the field of view reticle on the target and fire. The fire-control computer "corrected for range, lead, temperature, wind, munitions temperature, and barometric pressure," as Bing West describes the Marine Abrams killing fleeing Fedayeen fighters at Al Budayr from several hundred meters away.[47] In another instance, an Abrams tank crew had been monitoring a busy Al Budayr street filled with merchants and civilians some 650 meters away. Spotting four men with AK-47s amid the crowd, the machine gunner took aim through the day optical scope and killed one of the Iraqis with a single burst. After some minutes of scattering and confusion, the crowd returned, including the (now) three men with guns, upon which the tank gunner shot and killed another one. After figuring out where the shots had come from, the civilians started waving and giving thumbs up to the tank. The two men with guns did not reemerge. "The townspeople were so in awe of American technology," West writes, "that they believed the Marines could reach out and kill one person in their midst without harming others." He's quick to point out that Marines aren't always so precise: "War is a sledgehammer, not a scalpel."[48] But West's account illustrates the imbalance of American power gained through high-tech optics of combat.

West's Al Budayr account also suggests a certain redundancy in U.S. firepower. Abrams tanks are largely designed for killing other tanks. And yet at Al Budayr, and in many other engagements with enemy troops, the tanks were the first to fire, and often most decisively so. "The long-distance accuracy of the tanks miffed the Marine riflemen," as West explains so well, "who ran forward with their helmets and flak vests carrying ammo and water, sweating and covered with dust, hoping for just one clean shot, hearing the *bang bang* of the [Abrams] .50s [machine guns]." Marine riflemen started calling the recurrence of the Abrams eliminating the enemy threat before they got in range the "Oh Fuck Drill."[49] Another impressive weapons platform prominent in the ground invasion was the eight-wheel Light Armored Vehicle. With a crew of three, plus a four-riflemen (or scout) fire team, its armament included a Bushmaster 25 mm chain gun and two machine guns. These weapons' day and thermal sights—while not as powerful as those on the tanks—could target vehicles at two thousand meters. Close to Baghdad, at Al Aziziyah in early April 2003, Warrant Officer Mike Musselman spotted, through the sight of the chain gun, several Iraqi soldiers removing their green uniforms. One Iraqi fired his AK47 at the LAV and then started to run away. The Iraqi soldier turned "misty" after being hit by rounds from the 25 mm gun, as Musselman thought how crazy it was for the man to have shot.[50]

Even the grunts of the Marine infantry who fought on foot with standard-issue service rifles benefited from a myriad of battlefield technologies that allowed them to see farther and with greater clarity than their Iraqi counterparts:

- *Satellite technology.* While not in the hands of a Marine rifleman, or directly enhancing his vision, satellite-provided services had become an extremely practical and diffuse tactical asset by the time of the Iraq War. Ground force command and control functions relied heavily on GPS and other satellite technologies for navigation, situational awareness, and communications. Such technological superiority improved the positioning, perspective, and target acquisition of Marine infantry.[51]
- *Blue Force Tracker.* Depending on a satellite link through a GPS receiver, Blue Force Tracker pictorially represented the real-time location of friendly forces (and to some extent enemy positions) arrayed on the battlefield. Mapping software on laptop computers made Blue Force Tracker an essential tool for ground commanders as well team leaders in Humvees and other vehicles. With fratricide the leading cause of American casualties in the Gulf War, Blue Force Tracker is another excellent example of the practical differences that optical technology made in early twenty-first-century U.S. force projection.[52]
- *Dragon Eye.* A small model-airplane sized UAV (unpiloted aerial vehicle), the Dragon Eye gave battalion and company commanders a bird's-eye view of roadways, bridges, and enemy positions. Flying at ninety meters for up to an hour, Dragon Eye was remote controlled from its launch site and would send back live video images from its internal camera to the commander's monitor. Superseding the usual command and control hierarchy, the system provided ground units with immediate tactical intelligence—information that, again, informed the perspective of individual riflemen.[53]
- *Advanced Combat Optical Gunsight (ACOG).* A rifle combat optic for the M-16, one ACOG was initially issued to each twelve-marine infantry squad. ACOG is a fixed 4x optimal aiming scope that provided enhanced target identification and hit probability out to eight hundred meters. Its dual illumination feature allowed marines to keep both eyes open while engaging targets. Lieutenant Colonel Willard A. Buhl, battalion commander of the 3/1 in Operation Iraqi Freedom, after noting the complete superiority in marksmanship his marines enjoyed over Iraqi troops, singled out ACOG as an important factor, further describing it as a "force multiplier."[54]
- *Night Vision Goggles.* Amplifying ambient light from 30,000 to 50,000x, the third-generation night vision goggles issued to marines in Iraq featured an improved image-intensifier tube utilizing a microchannel plate resulting in a much brighter image than earlier models. "Night became day" is one account of marines' use of night vision goggles in Iraq. While something of an exagger-

32   *Light It Up*

ation, they did provide considerable tactical advantage; one could see, target, and shoot an enemy soldier two hundred meters away in the dead of night. "Night became lime-green and alive through the lenses" is another description of putting on the goggles while on patrol in Iraq: three houses down "I could see a cat prancing along a brick wall, the cat eyeing us as we eyed it"; next to the cat was "a Mercedes, it looked new, the finish gleamed in the daylight-bright glow of the night vision goggles."[55]

- *AN/PEQ-2 target laser.* Otherwise called an Infrared Target Pointer Illuminator/Aiming Light, AN/PEQ-2 is a laser sight mounted on the M-16 and other small weapons. AN/PEQ-2 gave marines a highly focused beam of infrared energy for weapon aiming and an adjustable focus infrared beam for target illumination. Using both night vision goggles and the AN/PEQ-2 sight, Marine infantry had vastly superior night vision and targeting capability. "With the advent of new technologies," as one writer in *Leatherneck Magazine* put in later in the Iraq War, "Marines are now able to fight completely in the infrared spectrum."[56]

Such optical technologies available to the Marine infantry when combined with its huge array of weaponry amounted to a magnitude of force projection that the Iraqis simply could not match. The marines knew that. Writing during the insurgency in 2008, Sergeant Michael Hanson spelled out the ongoing asymmetry on the ground. "Infantry Marines patrol with a squad consisting of 3 fire teams. Each fire team has four Marines with hand grenades, one M249 Squad Automatic Weapon, one M203 40 mm grenade launcher, and three M16s ideally with ACOG and PEQ2's. Each man has night vision capabilities and is wearing full body armor kit." After describing how "each fire team is its own maneuver element," Hanson asked: "If you were a 3–4 man insurgent cell would you attack a Marine rifle squad?" No, because you would "get killed fast." The Iraqis did "so in the past," though, and while they have since learned from that experience, the spring 2003 invasion remains an abject lesson in how impossible it was for them to match the Marines in conventional warfare.[57] As Lieutenant General James T. Conway, commander of the I Marine Expeditionary Force, put it to his troops in the opening day of the war: "This isn't a fair fight. We didn't intend it to be."[58]

In returning again to Keefe, we can square the overwhelming superiority the Marines enjoyed in optics and firepower with his focus on the Iraqis' poor marksmanship by concentrating on the particular perspective the sergeant had on and in the fighting. Keefe's experience of the enemy's bad shooting

Shock and Awe and Air Power   33

Figure 1.1. Sergeant Colin Keefe, Third Battalion, Fifth Regiment, April 2003, Sadr City, Iraq, on top of a CAAT (Combined Anti-Armor Attack Team) Humvee and next to a MK-19 grenade launcher. "Keep your eyes out" and "shoot the enemy before they shoot me" were Keefe's basic orders during the invasion. Keefe fought from the Humvee without armor or shields. (Colin Keefe)

resulted partially from the shock and awe magnitude of Operation Iraqi Freedom. Keefe personally saw little of the Third Marine Aircraft Wing, though. "I never saw any fixed wing (jets) engage in any sort of close air support," he recalls, "or do anything at all other than zip over our heads on their way to somewhere else." And Keefe's point of view benefited little from advanced optics: "My MK-19 just had old-school iron sights."[59] His unit's first rule of engagement during the invasion—"Keep your eyes out"—betrays the elementary nature of the war for Keefe. "If you're fired upon," he elaborates, "it's usually shoot back first and ask questions later." As part of a CAAT "ambush" team, Keefe fought from a Humvee—highly mobile but with no armor. "I stood up in the Hummer with my upper body sticking out. I didn't have any of these shields you see on them these days, I was just naked to the world," Keefe describes (fig. 1.1). "That was my job," he adds almost incredulously, "that's what I did." Having crossed over the berm from Kuwait into Iraq—"going forward into the unknown"—Keefe had little more between him and enemy troops than his naked eye and a MK-19 grenade launcher.[60] No wonder his combat memories concentrate on marksmanship.

In Keefe's case at least, the eye of a well-trained marine and the MK-19 proved to be a deadly combination. The MK-19 is a formidable weapon. With a firing rate of sixty rounds per minute, it has a maximum effective range of fifteen hundred yards, almost a mile. And, at the same time, "it's actually pretty devastating in close quarters," as Keefe found, "because it's fairly accurate." Designed to stop armored vehicles—"it will kill anything short of a full-on tank, and won't do a tank any good"—the MK-19's 40 mm grenades can penetrate two inches of armor plating and six inches of concrete. Keefe adds, "It'll blow a good sized hole in the wall of Iraqi houses." What it does to a human body is all the more devastating.[61] "If it hits you square in the chest," at close range, "it will blow your chest all out," Keefe describes. "If it hits you in a limb," he continues, "it'll blow it off." Keefe knows of what he speaks from the increasing number of firefights the 3/5 had as it neared Baghdad. An unusually extensive encounter took place some forty miles southeast of the Iraqi capital at Al Aziziyah, on the north bank of the Tigris, against the Al Maeda Republican Guard division, or its remains. "They'd been hurt pretty bad by air," Keefe recounts. While riding into Al Aziziyah, the Marines passed unarmed Iraqi troops walking the other way—which "was fine with us," as Keefe adds. "But there were still some hardcore guys holed up in this town," he continues, "and we go in and start duking it out with them." For Keefe the fighting started when, "as always," his CAAT "gets ambushed." Keefe details what happened:

> In one of the more dangerous for me personally moments of the war, there's two snipers up about, I don't know, 150 yards away, and on the second floor of a house and we're kind of pushing through on one of the main roads. We're obvious targets, and I'm sticking up, you know, out of the Hummer, so I'm an obvious target, and [after] I shoot a man, I can hear a bullet whiz by me. . . . I'm trying to find this sniper and he shoots at me again, and it's right by my ear. You can tell from the sounds where it's coming from to a degree, so I look out and I got a general area, so I'm just looking and looking and looking. He shoots at me again and misses me again, I can hear it just crack, right by me, but I see a muzzle flash that time. You know, so I'm not waiting for anything at this point, I pretty much yell out, I got him, and I just start laying into the entire top of this building with another couple of machine gunners. . . . We just demolished the top of the building, so you know, that was that.[62]

Keefe's combat memories tend to follow a pattern, a cognitive and emotional sequence beginning with vigilant, "eye-out" attention as his CAAT moves forward without knowing what lay ahead; clear recognition of his exposure to danger combined with sudden attack; a brief period of confusion or uncertainty followed by adjustment and response; and then satisfaction over winning the firefight, preceded by intense and skillful application of overwhelming force. After-the-fact incredulity over the enemy's bad shooting often accompanies Keefe's stories. One of the snipers he killed at Al Aziziyah had a "top of the line Russian sniper rifle," Keefe says: "I mean the fact that he missed me three times is bizarre." Each account's high point comes, perhaps unsurprisingly, in Keefe's own shooting, as in "I just start laying into the top of this building."[63] There's an almost ecstatic quality in how he remembers working the MK-19 and seeing its destruction.

Such climax amid a particular combat memory comes through in Keefe's account of his unit's next firefight against "foreign Islamic jihadists." His Humvee was once again ambushed when eight foreign fighters "pop out" of a tunnel. "They've got AK-47s and they're just wailing away and they're ten yards in front of us," Keefe recounts: "So, I think, you know, I'm done at this point, and I'm like holy crap." He didn't get hit, however, and, after the marine driving the Humvee spins it around to give Keefe a better shot, "I just lay into them with the MK-19," hitting two of the fighters. After the driver moved the Humvee to one end of the tunnel, Keefe spotted the fighters "dragging . . . a body or somebody up the tunnel that I'd hit." One of them "sees me and I see him, and he pulled his rifle up and I bring my machine gun around and we both just start shooting at each other, and I get him before he gets me." The Humvee driver then found another tunnel holding several enemy fighters. "I just start shooting grenades right up the tunnel, just hooked up the whole length of the tunnel," as Keefe describes the climax of this firefight: "I'm just lighting it up."[64]

What usually goes missing from Keefe's firefight stories is the death and destruction resulting from his rounds. That's one function of an expression like "light it up." It captures, in Keefe's words, the "rush" and "exhilaration" of manning such a powerful weapon as the MK-19. The expression goes to the optical quality of that rush and exhilaration, while maintaining the shooter's remove from the human casualty that occurs when marines successfully focus bright "light" on a target. That's not to say that Keefe has forgotten the death he caused. It comes out in the beginning and end of his Iraq War accounts and the "guilt" he feels from "killing so many people." Images of the dead and dying have certainly stayed with him. After the tunnel firefight, Keefe and his squad tried to help take care of wounded marines and enemy fighters. The brother of one of the fighters that Keefe wounded was on hand—distraught— watching his brother die before him as Keefe stood over the two of them. "That whole scene there," Keefe says, "I have nightmares about that." After combat, discharge from the marines, and developing a civilian life full with family and career, Keefe now "empathizes with people on the other end of my rounds." He elaborates: "It's not like they're necessarily bad guys . . . a lot of them are just people who grew up and enlisted in their own national army just like I did and are now defending their country against an invading army." Keefe feels "bad about people on our side that got killed," but he feels "a hundred times worse about the other people, that's what affects me later on down the road."[65]

Empathy is something that can be seen as antithetical to war and killing. The enemy, oftentimes purposely dehumanized in wartime, becomes—through empathy—someone, a fellow human being. To emphasize, though, Colin Keefe's empathy for his counterparts only came in Pennsylvania, well after the fact of the early spring 2003 shock and awe campaign. Keefe spoke of how a few of the toughest hard-bitten marines that he knew had broken down in battle; empathy for the enemy along with concern for self-preservation may have undone their capacity to fight. Most of the infantry in the I Marine Expeditionary Force acted like Keefe and in extremely close accord with Corps culture, doctrine, and training. Moreover, most of the death and destruction inflicted upon Iraq, its armed forces, and its civilians during the invasion happened from high above the ground—from the immediate and complete supremacy the United States enjoyed in air power. Empathy at such distance rarely happens, which has been no small factor in air power's prominence in U.S. force projection over the past several decades. Air power has also routinely betrayed the limitations of U.S. military force, as in the Iraq invasion, when, combined with lack of post-Saddam planning, the indiscriminate killing of shock and awe led to an insurgency that would keep the Marines in the war for years to come.

CHAPTER TWO

Of War Porn and Pleasure in Killing

How does the United States induce soldiers and marines to leave their homes, travel across an ocean, land on foreign soil, and, while risking one's own life, try to take the life of a stranger? Pulling the trigger has proved to be far from automatic. "Soldiers can never be transformed into mere instruments of war," as political theorist Michael Walzer has written: "The trigger is always part of the gun, not part of the man. If they are not machines that can just be turned off, they are also not machines that can be just turned on. Trained to obey 'without hesitation,' they remain nevertheless capable of hesitating."[1] Distancing the killers from their targets, as examined in the preceding chapter, has been one overarching response to this pressing problem of American empire. High-altitude bombing and, with the post–September 11 wars, predator drones have buffered Americans from the mortal dangers of battle, even as such remote-controlled force sharply increases civilian casualties.

A second response to hesitation over killing has moved in the opposite direction. Alongside the technological enterprise of separating the killers from the killed has been the equally vast cultural project of closing in on the action of war, capturing its violence, and aggrandizing its destructiveness and passions. If warfighters cannot "just be turned on" to kill in battle, perhaps they can be prompted to do so by reading, hearing, and *seeing* how others have done it before them. War literature has played a critical role in this representational dynamic. Human history, as the Harvard philosopher and psychologist William James recognized in 1910, has been a continual "bath of blood"; and appeals to war's "horrors" and "thrills" are written into the very foundation of Western culture. "The *Iliad* is one long recital of how Diomedes and Ajax, Sarpedon and Hector *killed*," James emphasized: "No detail of the wounds they made is spared us, and the Greek mind fed upon the story."[2] Shakespeare, Tolstoy, and Hemingway, among countless others, would continue where Homer left off, with the Iraq War producing a veritable cascade of contemporaneous literary renderings of battle and the hardship, excitement, adventure, heroism, tragedy, and loss that have always gone with it.

When it comes to motivating wartime killing, though, the moving picture has proved to be an especially effective representational form. Film theory has long recognized that both photography and cinematography enjoy a quality of *indexical* representation, a point-for-point correspondence between the object before the camera lens and the recorded image.[3] The mechanically or digitally determined verisimilitude of still and motion pictures has certainly played to the age-old compulsion to "see what war is like." But moving pictures do more than still images. In representing action—or "activity coming into being and being," as one film scholar puts it—moving pictures more fully capture behavior in combat. Point-of-view shots record the moving vision and changing sight lines of the warfighter. Compared to photographs, which tend to freeze frames of memory, film goes to imagination, insinuating itself into preconsciousness and into the very part of the mind that generates bodily movement.[4] Since the advent of cinema in the late nineteenth century, war has enjoyed a uniquely symbiotic relationship with moving pictures. War gives Hollywood one of its greatest and everlasting themes, while movies have played before several generations' eyes the perceptions of fighting and killing in combat.

Like no other medium, film has created an eye for battle. Since World War I, American soldiers and marines have conflated war and film, regularly explaining, perceiving, and *experiencing* combat via the war films they had seen earlier in their lives. In an oft-quoted example, World War II GIs advancing to forward battle lines in Belgium were heard to say, "We're off to the movies." Philip Caputo, a Vietnam War Marine lieutenant, has written that his many years of watching war movies as a youth made it easier to kill Viet Cong: "One part of me was doing something while the other part watched from a distance." Fast forward to the early twenty-first century, when comments equating actual combat with war films have been pervasive among marines fighting in Iraq. Sergeant Timothy Connors, for instance, in describing how he helped get his squad out of danger during the battle of Fallujah, explained that he coordinated fire "like a movie scene." In so many examples from the Iraq War, the comparison to film seems reflexive, almost natural, embedded in marines' perception. "When we crossed the line of departure in the tracks," Private Andrew Sean Stokes recalled the battle of Fallujah's opening, ".50 cals were going, tanks firing into the city, lots of stuff was blowing up—it was loud as shit. It was raining that night, it was dark and ominous, the perfect cinematography for a movie."[5]

The U.S. Marine Corps itself has been a great producer, collector, and purveyor of moving pictures about its ways of war. With its own film division and archives, the Corps has shot a tremendous amount of action footage. It took, for instance, more than 7 million feet of film during World War II, with 1,633,500 feet of it archived for internal use, Corps advertising and recruitment, and as source material for movies and television programming.[6] The Iraq War saw the deployment of combat camera teams made up of a photographer, a videographer, and a multimedia specialist.[7] Officially taken imagery of the Marines in Iraq was then collected, archived, and distributed by the Defense Video and Imagery Distribution System—an Atlanta-based military-civilian operation coordinating media for each of the U.S. armed forces.[8]

In addition to its own effort, the Marine Corps has greatly benefited over the past century from an especially close working relationship with Hollywood. Beginning with such films as *What Price Glory* (1926) and *Tell It to the Marines* (1926), the Corps realized that Hollywood studio productions would contribute immeasurably to its constant public relations campaign, from securing its place among the other armed forces, to gaining congressional funding, and to meeting its recruitment goals. In exchange, the Corps opened its operations to countless production teams from the entertainment industry. Camp Pendleton, the large Marine base in southern California, has been used by major studios for producing feature-length combat films. While gaining military expertise and use of equipment and personnel, the studios physically adapted the base to meet historical specifications—from a beach on a Pacific island for Republic Pictures' *Sands of Iwo Jima* (1949) to snowy Korean mountains in Warner Brothers' *Retreat Hell!* (1952). The Marines would then use these mockup terrains in preparing for future military operations, as war making and film production continued to influence each other. As part of its Division of Public Affairs, the Corps has long assigned a liaison officer to Hollywood, usually a lieutenant colonel whose job includes the protection and promotion of the Marines' image. *Battle Los Angeles* (2011) is just one recent example of what results from such close collaboration: a science fiction film about an alien invasion, it amounts to an elaborate "product placement" for the Corps in featuring the Second Battalion, Fifth Regiment's heroic response to the attack. "This is a movie about Marines"—the film's star Aaron Eckhart told a crowd at the Camp Pendleton premiere—"kicking ass."[9]

With the advent of television, the Marines found a whole new medium for self-promotion. CBS's situation comedy series *Gomer Pyle U.S.M.C.* (1964–69) is *sui generis*. Shot at Camp Pendleton with the Corps' full cooperation, and aired at the height of the Vietnam War, the highly popular series merits its own critical monograph. In a more systematic vein, the Marine Corps employs Madison Avenue advertising agencies to produce TV recruitment commercials often played during weekend football games and other sports events. Some of these spots—such as "Pride of the Nation" (2005)—incorporated archived footage taken during combat operations. The Marines and the Navy used high-definition digital images from the Afghanistan War in making "Enduring Freedom: The Opening Chapter" (2003), the controversial five-minute promotional film played as a trailer in hundreds of U.S. multiplexes during the lead-up to the Iraq War. Even though many theater complexes stopped playing "Enduring Freedom" because of its propagandistic purpose, the Marines and the Navy collaborated on a sequel, "Iraqi Freedom—Chapter II" (2003).[10] As the Iraq and Afghanistan wars continued, the Marines worked closely with commercial TV, Hollywood, and the full gambit of the military-entertainment complex to advance its image and interests at home and abroad.

The Iraq War happened in concert with dramatic changes in digital media, an evolutionary turn that brought rapid advances in the synchronic capturing and sharing of moving pictures with affordable video cameras and smart phones and through the internet. American marines and soldiers in Iraq, armed with digital cameras (some on their helmets), shot footage of their fighting and its results, while editing software allowed them to piece together films from camp, often overlaid with heavy metal soundtracks.[11] So began "war porn."[12] The genre would flourish with the 2005 founding of YouTube, where American-shot videos have been joined by competing works filmed by mujahedeen, Jihadist, and insurgent fighters, while other video-sharing sites would more overtly support anti-American material. Some have dubbed it the "YouTube War," a propaganda battle of perception fought out with viral images on the web for the hearts, minds, and eyes of a global viewing public.[13]

This chapter considers how war porn from Iraq has helped project U.S. military force by concentrating on the genre's similarities to sexual pornography. The video material's illicit nature certainly supports the pornographic label. "The soldiers use the word," as defense industry analyst P. W. Singer comments, "because they know there's something wrong with it."[14] More specifically, though, "war porn" coheres because the combat videos create desire to carry out the behavior they represent. It may be wrong—a guilty pleasure—to trade and revel in such pictures; but given the violence of the fighting itself, the videos' impropriety seems secondary. War porn's power comes in motivating further killing in battle. An equally descriptive (if longer-winded) label for war porn is "pump-you-up-to-kill-the-bad-guys videos."[15] War porn has a type of local and immediate propagandistic effect: it advances the interests of the state by making familiar and even enjoyable the degradation and destruction of human life targeted by the military. At the same time, though, as a product of personal media, war porn almost precedes ideology in its vernacular origins and appeal. Combat videos on YouTube need to be understood within the larger visual cultural context of the early twenty-first century, when satellite and cable television fixed on the entertainment value of the Iraq War, and when U.S. society as a whole had become more militarized than any time since World War II.

Pornography Is the Theory, and Killing the Practice

In the final scene of the HBO miniseries *Generation Kill* (2008), a group of Reconnaissance marines huddles around a laptop inside an empty Baghdad warehouse to watch a video highlighting the past several days of invasion and combat. Young and proud, as one might expect, the marines are enraptured with the moving pictures of their exploits and themselves "in country"—mobile, deadly, and apparently having the time of their lives. Shots of Cobra helicopter rocket attacks and exploding buildings are intercut with grunts mugging for the camera and the platoon posing for a triumphant photograph with a large American flag beneath a Saddam statue. But as the images turn to Iraqi dead, and as the marines start to take in a broader view of Operation Iraqi Freedom, their visage changes from lustful exuberance to, as literary historian Stacey Peebles describes, "somber resignation. One by one, they walk away from the show." It's a powerful scene, one that can be read to encapsulate in four minutes and thirty-nine seconds the long arc of many veterans' perception and memory of war: from ignorant and deadly enthusiasm to wounded and often guilty disillusionment. For Peebles the scene illustrates the necessary disjuncture between war and its filmic representation. "The experience of real war," she concludes, "always exists outside of the frame."[16] Very much open to interpretation, however, the scene also demonstrates film's determination of war making. Its first two minutes amount to a study in war porn. Awash in the glow from the screen, the marines' faces are transfixed by sequential going-to-war images, their expression changing only when registering pleasure over the explosions, destruction, and the dead. The invasion video is exactly the kind of production that would become so common on YouTube, where such a totem to the unit's prowess would elicit desire for similar accomplishment in marines to be.

Most definitions of pornography suggest a causal relationship between sexually explicit text and sexual behavior—that is, not only does pornography represent human sexuality in exaggerated, lewd, or degrading ways, but such portrayals are produced and consumed for the purpose of prompting, endorsing, or recommending sexual activity of the same nature.[17] Feminist opposition to pornography concentrates on cause and effect. Radical feminist Robin Morgan's well-known statement that "pornography is the theory, and rape the practice" draws out the point.[18] Her contention is that rape and other sexualized violence against women is the likely outcome, the intended consequence, of an overarching system that portrays and treats women as inferior subjugated beings and as objects. Pornography is the abstraction of men's sexual domination and abuse of women. One need not wholly endorse this position, or even weigh in on the complex issue of pornography, to appreciate the war porn analogy. Killing in combat is the likely outcome, the intended consequence, of the already highly militarized early twenty-first-century U.S. culture that further developed the free and accessible flow of combat videos taken from recent battles and edited and circulated to celebrate the action and accomplishment of warfighting. If pornography seeks sexual arousal and release through explicit representation of human sexuality, then war pornography seeks similar arousal of passions toward violence and killing and release through explicit representation of martial combat.

The war porn analogy works so well because, like sex, the experience of war contains and elicits pleasure. Testimony to that fact is legion, resonant since antiquity, and mounting among Americans in combat since the Civil War, with those fighting in the U.S. Marine Corps finding and expressing more than their fair share of it.[19] Pleasure in war occurs via the eye and the perception of seeing. Iraq War YouTube videos attest, for instance, to the pleasure of watching with night vision goggles an Apache or Cobra helicopter attack a ground target in darkness, its pale green tracer rounds spilling down in a rainbow-like arc, eliciting "oohs" and "aahs" like Fourth of July fireworks. Some pleasures of seeing are more immediately sadistic. In his book *Generation Kill* (2004), upon which the HBO series is based, Evan Wright remembers a drunken First

Reconnaissance marine saying in Baghdad, "You know I've fired 203-grenade rounds into windows, through a door once. But the thing I wish I'd seen—I wish I could have seen a grenade go into someone's body and blow it up."[20] Pleasure also comes in killing and the action of combat itself, although almost always mediated in ocular perception and expression. Marine Reconnaissance first lieutenant Nathan Fick speaks of the mounting pleasure he found in Iraq. "Now, shooting grenades at strangers in an unnamed town," Fick writes in his memoir *One Bullet Away: The Making of a Marine Officer* (2005), "I was kind of enjoying myself. The long sought hyperclarity had kicked in. I saw a young man crouching in an alley. He wore dark trousers and a blue shirt. . . . I lobbed a grenade at him and the round exploded above his head. I watched him fall."[21] Later, after several more firefights, Fick recounts combat's mounting pleasure of sport and play:

> I saw in the platoon a glimmer of something I was starting to feel in myself: excitement. The adrenaline rush of combat and the heady thrill of being the law were addicting to us. This was becoming a game. I was starting to look forward to missions and firefights in the way I might savor pickup football or playing baseball. There was excitement, teamwork, common purpose, and the chance to demonstrate skill. I didn't have the luxury of much time for reflections, but I was aware enough to be concerned that I was starting to enjoy it.[22]

Fick comes close to admonishing himself for these feelings, avoiding the revelry in killing that marines fighting with him often expressed and for which the Corps has become renowned. The most infamous statement during the Global War on Terror regarding the fun of killing came from Marine Corps lieutenant general James Mattis, who said in 2005 that "it's a hell of a lot of fun to shoot them [Afghani men]. Actually it's quite fun to fight them, you know. It's a hell of a hoot."[23] While unusual for a high-ranking officer to make such a statement in public, its message is downright ubiquitous in the popular culture surrounding the Marines and U.S. warfighting overall—from G.I. Joe action figures to *Fightin' Marines* comic books, and from World War II Hollywood combat movies to war porn videos from the Iraq War.

Having noted the pleasure principle embedded within modern American military culture, what can be said about the actual effect that war porn videos have on warfighting? Most scholars who have addressed the relationship between combat films and killing in combat do not venture direct causality between the two. In his work, *American Samurai: Myth, Imagination, and the Conduct* of Battle in the First Marine Division, 1941–1951 (1994), Craig M. Cameron discussed the close synchronization between Hollywood and the Marine Corps during World War II, but went only as far to say that the resultant movies were "vehicles for expounding and perpetuating the [Corps'] mythic dimensions" of battle prowess. Lieutenant Colonel Dave Grossman's influential book *On Killing: The Psychological Cost of Learning to Kill in War and Society* (1995) advances a forceful critique of combat video games as "killing enablers" and "murder simulators"; he's less direct, though, regarding the effect of combat films, explaining that moving pictures "desensitize" viewers to violence, making them, in turn, more likely to commit it themselves. Joanna Bourke may be the most careful in distancing herself from the feminist antipornography position of causality, insisting repeatedly in her book *An Intimate History of Killing: Face to Face Killing in 20th Century Warfare* (1999) that a "direct relationship" does not exist between war films and "the urge to kill"; while combat films "do not directly stimulate imitation, the excitement they generate creates an imaginary arena packed with murderous potential and provides a linguistic structure within which aggressive behaviour might legitimately be fantasized."[24] In Bourke's formulation, combat films construct an emotional and cognitive context more conducive to expressions of wartime violence—a context of fantasy, imagination, and desire for killing.

This combination of combat films eliciting desire for violence and warfighters registering pleasure in combat, with some of those marines and soldiers relating and experiencing that pleasure through films they saw before battle, could be said to create a causal loop between culture and war. While possible to identify the impact of specific films on individuals in combat, it's more critical to recognize that, in general, films and other cultural texts *inform* perception and behavior in war, which, in turn, through its representation in memoirs, novels, blogs, Hollywood films, and war porn videos, make up the myriad cultural texts that will inform the next generation of warfighters. And that process of informing—that creation of meaning—has largely happened within the bounds of the hypermasculinist imagination: a culturally determined context in which killing in battle, like sexuality, is commodified within an economy of scarcity as an eternally sought after experience that must be taken when rare occasion arises. "Get some," as marines say.

YouTube war porn videos are particularly effective in how they tighten the loop between the conduct of battle on foreign ground and its representation, with very little time between behavior and cultural production based on that behavior and then its consumption. Optical conventions structure expectations of what war looks like to viewers of combat footage. While those optical conventions are diffuse in American culture at large, some of the viewers inevitably become combat soldiers and marines who go to battle with "what war looks like" in their heads, which, in turn, shapes the experience of battle. Anthropologist Jennifer Terry has commented that Jean Baudrillard's "once seemingly radical observation that the simulacrum has attained the status of the real in post-modern culture, sounds quaint and understated in this contemporary mediascape of war and viral video."[25] We know that war porn has been shown in Marine boot camp and played among troops deployed in Iraq before combat, patrols, and other military operations; the fact that marines decide to watch the videos in such close proximity to that behavior creates a de facto causal relationship between it and those texts; viewer comments on YouTube divulge the subjective effect of combat videos, including their capacity to elicit pleasure and to prompt interest in going to war.

Classic Hollywood Combat Films

Before the advent of YouTube, before the internet, before combat video games, and before digital media altogether, Hollywood combat films had the greatest pornographic effect in American war culture. To review that genre and consider its allure through the mid- to late twentieth century is to come to a closer understanding of the relationship between moving pictures and motivation to fight and kill in combat. Here, World War II combat films have enjoyed unusual power. With the genre itself born during the "good war," and with the pastime of moviegoing as popular as ever in the postwar years, the heroic combat of World War II—what some film historians call "Hollywood's military Ur-narrative"—became a formative visual experience for many male members of the baby boom generation.[26] Happening on big screens during Saturday matinees as well as on late-night TV, this extraordinary cultural transference from one generation to the next is well documented in American Vietnam War memoirs. "I wanted the romance of war, bayonet charges, and desperate battles against impossible odds," former Marine lieutenant Philip Caputo spoke of going into the Vietnam War: "I wanted the sort of thing I had seen in *Guadalcanal Diary* and *Retreat, Hell!* and a score of other movies." World War II's on-screen messages included both an eagerness to enlist and an enthusiasm for over-the-top heroics in battle, oftentimes tagged as "going John Wayne" in derisive commentary on Hollywood's distortions. "I keep thinking about all the kids who got wiped out for good," correspondent Michael Herr wrote in *Dispatches* (1977): "We'd all seen too many movies, stayed too long in Television City."[27]

The World War II combat movie, as film historian Jeanine Bassinger has charted, adopted the twin goals of realism and propaganda.[28] "This is the factual record of the Second Marine Raider Battalion," an opening frame of printed words proclaims in *Gung Ho* (1943), "from its inception seven weeks after Pearl Harbor, through its first brilliant victory." The Marine Corps provided technical advisers, uniformed extras, and equipment for *Gung Ho*—just as it did for *Wake Island* (1942), *Guadalcanal Diary* (1943), and *Sands of Iwo Jima* (1949). The screenplay for *Wake Island* came "from the records of the USMC," the film's opening declares, its "action . . . recorded as accurately and factually as possible." As for propaganda, *Wake Island* explains that the marines sacrificed their lives defending that outpost in the south Pacific "because in dying they gave eternal life to the ideas for which they died." A third pornographic component of pleasure seeking comes through in *Wake Island*'s rich rendering of Marine firepower and the taking of Japanese lives. With the enemy's warships firing on the island while approaching for invasion, the Marine guns are hidden and stay silent. American positions take a beating. Finally, as Japanese landing parties are lowered into the water, the marines are ordered to "Open Fire." They do so with a gleam in their eyes. The unwitting Japanese troops drop en masse as the Marines' Hymn ("From the Halls of Montezuma . . . ") plays through the fighting, melding Corps spirit with satisfaction through killing. *New York Times* film critic Bosley Crowther watched the newly released movie at Quantico Marine base in August 1942 with "some 2000 fighters" preparing for war in the Pacific. "Those Marines have a personal interest involved," Crowther recognized, reporting that they "cheered" the film "with thunderous applause."[29]

Taking pleasure in killing is at the center of *Gung Ho!* Released a year after *Wake Island*, it dramatizes the August 1942 attack on Japanese forces on Makin Island, one of the first offensive actions by Americans in the Pacific. *Gung Ho!* opens with the Marines' standard for the handpicked Second Raider Battalion: desire and capacity for "close combat with the enemy." Officers select the men by asking, "Why do you want to kill Japs?" They "want killers"—those who like "the feel of the blade." There's an explicit promise to the battalion, and the viewers, that "we're going to see some fun before we're through." And the film doesn't disappoint. *Gung Ho!* is exceptional, even within the combat film genre, for its "sudden and concentrated action," as *Times* critic Crowther put it in 1943. "Without any subtleties, the performances are all rugged," he continued euphemistically, as the movie's second half fills the screen with sacrificial charges against Japanese machine gun nests, hand-to-hand combat, eye gouging, and killing with knives. "The stabbing and the sticking," Crowther commented, "go on ad nauseam." But, overall, his review is positive and approving. "The settings are true," Crowther wrote in praising the film's realism. And "the fighting on the Island," he concluded with unmistakable satisfaction, "is as hot and lurid as any that we've seen."[30]

Pleasure seeking through battle can also be found in *Sands of Iwo Jima*, probably the best-known World War II combat film, and one that helped pull a new generation of young American men into the Marines. Wanting *Sands of Iwo Jima* to crystallize its World War II legacy, the Marines worked intently with director Allan Dwan to instill both heroism and realism into this early Cold War production. Dwan consequently gave speaking parts to several veterans from the actual Pacific campaigns; he made liberal use of documentary footage for establishing shots and to carry the action of battle scenes; he even had the actors go through an abbreviated boot camp with a USMC drill sergeant.[31] Replayed countless times in the nation's theaters and on television during the 1950s and 60s, *Sands of Iwo Jima* projected the adventure and deadly romance of combat well into the Vietnam War era. It effectively joined two generations, parts of which would also oppose each other over the Vietnam War.

Ron Kovic, Vietnam War veteran and author of the memoir *Born on the Fourth of July* (1976), saw *Sands of Iwo Jima* in the mid-1950s at a Saturday matinee with his childhood friends in Levittown, New York. They watched John Wayne, playing the rugged Sergeant Stryker, lead his squad onto the beaches of Tarawa and Iwo Jima. "We sat glued to our seats," Kovic recounted, as Styker's larger-than-life marines fought their way through line upon line of Japanese troops. "They showed the men raising the flag on Iwo Jima with the Marine hymn still playing," Kovic wrote, "and [we] cried in our seats." In Kovic's memory, the drama in *Sands of Iwo Jima* melded with playing war in the woods with "Pete and Kenny and Bobbie" after the show and also with his decision as a seventeen-year-old to join the Marines. The Vietnam War went badly for Kovic. Critically wounded, and permanently paralyzed from the waist down, Kovic's *Born on the Fourth of July* epitomizes the anguish and sense of betrayal and loss among some Vietnam War veterans: "I gave my dead dick for John Wayne!" In adapting the book into a 1989 film, Oliver Stone meant to dramatize Kovic's

48   *Light It Up*

lament. And so he did. Although Vietnam War films from the late 1970s and 80s extended the generational transference of pleasure through combat, despite many of the movies' intended criticism of war. William Broyles, writer and former U.S. Marine lieutenant during the Vietnam War, noted in 1984 that "last year in Grenada American boys charged into battle playing Wagner [following Francis Ford Coppola's *Apocalypse Now* (1979)], a new generation aping the movies of Vietnam the way we aped the movies of World II, learning nothing, remembering nothing."[32] As former marine Anthony Swofford has illustrated so convincingly, the most didactic and brutal of antiwar films still excite the would-be warrior.

With *Jarhead* (2003), Swofford's memoir of his experience as a marine sniper in the Gulf War, there's little doubt about the intimate influence of Hollywood on men preparing for battle. After receiving word in early 1991 of deployment to Kuwait to attack Iraqi forces, Swofford's unit rents as many combat films as it can find for holding a "Vietnam War Film Fest." He writes: "We sit in our rec room" and for three days "watch all of those damn movies, and we yell *Semper Fi* and we head-butt and beat the crap out of each other and we get off on various visions of carnage and violence and deceit, the raping and killing and pillaging." Along with Swofford's close description comes a more removed and critical take of what the films meant to him and his fellow marines: despite the "talk that many Vietnam films are antiwar," they "are all pro-war, no matter what the supposed message, what Kubrick, Coppola, or Stone intended"; he explains that "the magic brutality of the films celebrates the terrible and despicable beauty of [the marines'] fighting skill." Swofford makes it explicit: "Filmic images are pornography for the military man; with film you are stroking his cock, tickling his balls with the pink feather of history, getting him ready for his First Real Fuck." As if the point needs elaboration, he concludes: "Now is my time to step into the newest combat zone. And as a young man raised on the films of the Vietnam War . . . I want to . . . kill some Iraqi mother fuckers."[33] Never has the Marine eye for battle been more illicitly construed. And never has the pornographic effect of combat films been more plainly rendered.

It's from this context of desire for killing arising from combat movies and the reciprocal promise of finding pleasure in real war that war porn videos emerged in the early twenty-first century. Hollywood didn't stop making war films in the post–September 11 period. *Home of the Brave* (2006), *The Marine* (2006), *In the Valley of Elah* (2007), *Hurt Locker* (2008), *The Messenger* (2009), and *Green Zone* (2010) are among the many feature-length fictional movies on the Iraq War. One would have to return to World War II to find such prolific filmmaking on a contemporary war. And within the proliferation of new media war porn, the canon of American combat films would resonate, intertextually shaping the composition, interpretation, and meanings of the Iraq War videos. "Iraq: Bird Is the Word," uploaded to YouTube in 2006 and one of the most viewed videos of the genre, was self-consciously produced as an "ode to *Full Metal Jacket*"—an effort to get past the idea of "glory in killing" and show "the slaughterhouse for what it is."[34] Despite these intentions, many of the some 10,017 YouTube comments on "Bird Is the Word" document the video's pornographic effect. Along with the dozens of "Get Some" responses—that Marine utterance encapsulating the imbrication between aggressiveness in combat, sex, and pleasure—could be found the enigmatic lines excerpted from *Full Metal Jacket*: "Is that you John Wayne? Is it me?"[35] When first spoken, by the film's protagonist Joker, the words question whether *Sands of Iwo Jima*'s iconic Sergeant Stryker and his gung-ho fighting spirit have any place in a post–"Good War" world. Commentary on YouTube war porn is not without its critical distance, but Joker's irony may have been lost on many amid a web culture so saturated with enthusiasm for killing in war.

Marine Moto on YouTube

The short history of war porn often begins with the infamous gore video from the mid-1990s of Chechen rebels killing six Russian soldiers by slitting their throats. By the early twenty-first century, Islamist extremist groups routinely filmed bombings and other violent acts and then posted the videos on the internet for purposes of propaganda and recruitment. With the Afghanistan and Iraq wars, the U.S. Department of Defense slowly tried to match radical Islamic presence on the web; in March 2007, for instance, it launched its own YouTube-like channel called Multi-National Force-Iraq (MNFIRAQ). Featuring videos of the U.S. military dispensing humanitarian aid to Iraqi civilians, along with decidedly sterile or "clinical" views of combat, MNFIRAQ attempted to counter not only enemy material but also the proliferation of war porn videos made by its own troops.[36] Whether the countless videos showing U.S. marines and soldiers taking pleasure in fighting insurgents run against the long-term strategic interests of American war making is open to interpretation. My argument here is that American empire, to some extent, depends on the vernacular pleasure principle of combat. But such videos showing U.S. marines urinating on Afghani rebel corpses and an Apache helicopter gunning down an unarmed TV crew clearly hurt any effort to win the hearts and minds of those living under American occupation. Meanwhile, the tens of thousands of deployed troops with considerable in-camp downtime and with up-to-date digital cameras and computer software made it all but impossible to staunch the production, posting, and circulation of war porn videos.[37]

With some analysts of early twenty-first-century combat videos categorizing them as either hard-core or soft-core porn, it's best to place them on a continuum: on one end is the simple snuff film–like close up of a person being executed or killed in combat; and on the other end is the longer music video–like production—texts that certainly highlight violence but do so within something of a narrative arc or that at least contextualize that violence within a sequence of images often identifying a certain military operation and unit and that almost always use a heavy metal audio track to establish the exhilaration and pleasure of combat.[38] While hard-core turns on sensationalist effect—so *that's* what it looks like to get shot in the head—videos leaning toward soft core try to inspire, focusing on the action of warfighting and rarely showcasing its grisly results.

Marines call attempts to inspire *moto*: short for *motivation*, it refers to anyone or anything that promotes the hard-charging, risk-taking behavior that the Corps wants from its troops. Sometimes moto is used among marines almost critically or sarcastically to designate something they see through—as something they know is constructed to get them going. This is "clearly a MOTO vid," commented MarineCorpsMarksman on YouTube about "3rd Battalion, 1st Marines in Fallujah Iraq 2004" (2006): "Not really supposed to be a fair show of reality."[39] Further comments on the video demonstrate its effect. "SEMPER FI! Our boys are the best! they're gonna kick terrorists asses"; "OORAH! I'm thinking of joining the Marines"; and "the most popular music video of the entire iraq war, its [sic] been reposted countless times and each repost has thousands apon [sic] thousands of views, all i have to say is oo rah marines!!" Viewers posted these responses during the Iraq War, between 2007 and 2010.[40] More recently, in 2011–12, with U.S. troops withdrawn, the video continued to inspire: "I USED TO WATCH THIS BEFORE PATROL IN FALLUJAH, NOW I WATCH THIS BEFORE I TAKE AN EXAM IN COLLEGE !!"; "25 yr old former marine. shit makes me want to reenlist"; and "Marine Corps for me in 5–10 years. Following my brother who is 20 in 3rd Battalion. Oorah."[41]

Several YouTube comments on "3rd Battalion, 1st Marines Fallujah Iraq 2004" reported that the video had been played in boot camp during the war.[42]

While some older veteran marines feigned outrage over "videos in camp"—"What is my Corps coming to"—it's not difficult to understand why "3rd Battalion, 1st Marines" had been shown to recruits.[43] As discussed at length in the next chapter, urban warfare had been one of the greatest challenges to the U.S. Marines going into Iraq. And this five-minute text amounts to a "how to" video of urban combat, including clearly shot scenes of fire teams patrolling city streets and entering and blowing up houses; it also features the full panoply of Corps weaponry and firepower—from Abrams tanks and Cobra helicopters leveling buildings from afar, to the marine and his rifle in rooftop firefights. The video contains no gore, except for a very brief pan shot (at 2:55) of a dead Iraqi soldier.

The moto payoff in "3rd Battalion, 1st Marines" is its illustration of "marines doing what they do best," as a common YouTube comment puts it. "Cool," a German viewer responded in 2008 to the Fallujah video, "I wish I could play there too."[44] Its inspirational value is heightened by the soundtrack featuring the South African band Seether's post-grunge song, "Out of My Way." Before the song, at the beginning of the video, in synch with the image of a red pulsing light, the video excerpts *Full Metal Jacket*'s Gunnery Sergeant Hartman: "God has a hard on for marines. Because we kill everything we see." Next comes the voice over of two marines from the 3/1 anticipating the fighting in Fallujah: "This is what people join the Marine Corps to do. You might be in the Marine Corps for twenty years and never get this chance again—to take down a full fledged city full of insurgents. . . . If you want some, get some." A marine is then shown lacing up his boots and buttoning his camie. A battle flag with John Wayne's image on it flutters in the wind (at 0:55–1:08).

Not all of the posted commentary following "3rd Battalion, 1st Marines" and other YouTube war porn from Iraq shares in the genre's bloodlust. In fact, the verbal exchanges are impressively variegated, often substantive in their consideration of war for oil, the spurious claims of Iraq's weapons of mass destruction, and other issues revolving around military force and American empire. A comment made in 2008 on "3rd Battalion, 1st Marines" is emblematic: "Sure, wars can bring a country out of recession, but at the price of many people's lives. War is hell, and this war is fucking stupid. Unfortunately, we can't just leave Iraq, or the country will fall apart."[45] Much of the critical commentary focuses on the fact of killing—as represented in certain videos and in the Iraq War itself. Some postings go as far as trying to adjust the collective conscience of the video producers and viewers. In response to a 2011 video celebrating predator drones, a commenter points out the following year that one of its clips is out of place. It's actually from the infamous instance when "an Apache helicopter opens fire on a group of unarmed civilians and TV crew. And by the way," freechristiania added, "predator drones have been used in numerous attacks where civilians have been the victims. Wind down on your war horneyness."[46] Some comments adopt a seemingly unassailable voice of reason, as in Bee Burgers' late 2012 comment attempting to split the difference between those enjoying and those criticizing a video showing marines throwing a hand grenade into an Iraqi house:

> It's really screwed up the way people respond to this. On one side, we have those who think these Marines are sadists, as compared to their enemy who regularly hides in buildings such as these to ambush Marines and strap bombs to civilians.
>
> On the other side, you have those who think all Muslims should die, as though that kind of hatred will somehow solve anything.
>
> If you've responded in any way other than, "War sucks and it makes good people do bad things," you're a dumbass.[47]

Despite such efforts, the discourse emanating from YouTube war porn videos has been structured by several binary oppositions, a series of interlinking dualities between "true Americans" who accept and even welcome the massive killing that comes with war and the "commies," "hippies," and "liberal bastards" who don't; marines (and other military initiates) and the untested civilians whom they protect; or, put differently, "People [who] are out there putting there [sic] asses on the line" and those "sitting [on their] fat ass[es] behind a keyboard and bashing them."[48] In each of these formulations, the first position constitutes the martial-minded center of the YouTube war porn viewership; and within that core community, as Bee Burgers suggests, lies a most fundamental binary: human beings whose life and autonomy are presumptively respected and those "dune coon muslim terrorists" who should be killed.[49]

With *terrorist* the most common label within YouTube war porn discourse for the U.S. military's enemy and target in Iraq, that word has hardly been used discriminately, as it seems to cover retrospectively almost anyone caught in American line of fire. It's also been written in close tandem with such expansive slurs as "stinking Muslims," "goat herders," and "sand monkey towelhead motherfuckers." Dehumanization of the enemy is a long-standing strategy for making modern war. A less-than-human enemy soldier is easier to kill or "exterminate," as in the American metaphor for killing Japanese in World War II;

and dehumanization blurs the line between enemy combatant and civilian—or, rather, it helps the warfighter forget that line altogether.[50] As one homicidal refrain has it among U.S. marines commenting on YouTube Iraq War videos: "Kill them all, let god sort them out."[51] Such commentary objectifies Iraqis as an indistinguishable mass of foreign targets, while the videos themselves work pornographically, fetishizing the action of eliminating Orientalized others—even as the YouTube texts seldom concentrate on the piercing or tearing of flesh, or on the morbidity of death in its own right. Compared to the 2004 Nick Berg beheading, for instance, or the 2014 ISIS videos of American journalists' murders, American YouTube war porn from the Iraq War, while certainly denigrating an ill-defined enemy, is less overtly political: it's intended less to incite the enemy to further conflict than to motivate those at home and already in military service by seeing and enjoying combat from the warfighter's perspective.[52]

That's not to say that videos in the war porn genre aren't quite explicit in aggrandizing the destruction of an objectified enemy. "Marines Get Some: Die MF Die," for instance, uploaded to YouTube by MarineTroopSupport in 2007, contains the key components of Iraq war porn: introduction of a particular unit of combat troops, in this case, part of a Marine platoon (at 0:18); highlights of those marines in action, shooting their weapons, throwing grenades, taking down buildings, and watching tanks and artillery blow up buildings from a distance; a hard hitting sound track, Dope's heavy metal song "Die Mother Fucker Die"; and simple identification of the enemy, as unknowable other, along with the imperative, as summed up in the title track, that they be destroyed. After the video begins with the written heading "Get Some!!!!" (at 0:03), it fixes on a photograph of two enemy fighters, their faces covered by keffiyehs (at 0:09–0:13). The most violent action footage shows marines shooting Iraqis at close range in a narrow street or alleyway (at 1:22–1:26); while people scatter, some of them undoubtedly hit, the video doesn't linger on any single wounded or dead figure. The fetishization of the enemy in "Marines Get Some" comes in closely edited, rapidly shown stills of enemy combatants (e.g., at 2:40–2:42): each sequence includes eleven photographs, with each still synchronized with the syllables of the song's chorus: "Die mother fucker die mother fucker die." It works: "The reason that i love this video is because it shows us (the marines) doing what we do best. helping spineless bastards meet their maker since 1775! oohraw!" said one commenter in 2008. "Great video and very well edited!" added another the same year. "Plus," the YouTube commenter continued, "I like to see those terrorist motherfuckers get killed, its [sic] that simple."[53]

Violence against the U.S. enemy is more explicit in the war porn video "Bullet with a Name on It." Uploaded by WilcoUSMC in 2006 and set to Nonpoint's heavy metal song of the same name, "Bullet with a Name on It" is another canonical Marine-focused YouTube Iraq War action video, enjoying well over one million views and eliciting several thousand comments. A good portion of those comments concentrate on footage (at 0:43–1:07) of a rolling bus in flames, having been lit up, finally coming to a stop; a black-clad man then jumps out and, in a hail of bullets, runs for cover in a small hutlike structure; it, too, is "lit up," along with the man inside, whose corpse is shown lying twisted in a pool of his blood.

This scene in "Bullet with a Name on It" is one of pursuit, predation, and payoff. In response to comments suggesting that the bus had been civilian, and the man unarmed, the Marine Corps faithful among YouTube war porn viewers are unforgiving, exuberant, and absolute in the killing. "Kill or be killed," wrote one marine in 2006: "This is no fucking game of kindness and simpathy [sic]. Obey or die. We rule this land with a killing machine, so take your chances and we will send your raghead ass to fuck all the virgins you want." Commenting at about the same time, Marines137 asks, "Who gives a fuck if he was unarmed? They should put a bullet in his head anyway. Marines are born to kill! Semper-Fi OOH RAH!!!"[54] One senses that these comments came from marines deployed in Iraq, perhaps drawn from their sensibility, amid great danger, during patrol and combat—a sensibility that was then diffused into online culture and YouTube discourse and later triggered by the fantastic images of violence and killing in the video itself.[55] Such speculation points to the problematic nature of using viewer comments as a primary historical source: it's virtually impossible to know with certainty the identity of any one commentator or the extent of exaggeration and hyperbole. My analysis here rests on reviewing hundreds upon hundreds of YouTube comments on war porn videos, determining repetitive rhetorical strategies and fixations, and selecting representative examples for summary and interpretation.

A vivid example of the lurid attraction to the enemy's destruction can be found in the "Iraq War—Let the Bodies Hit the Floor" war porn video, uploaded in 2009 and set to Drowning Pool's extremely popular heavy metal song "Bodies." Third Battalion, First Regiment marines recall hearing the song played by Coalition Psychological Operations in combat during the second battle of Fallujah:

Driven by hate consumed by fear
Let the bodies hit the floor
Let the bodies hit the floor[56]

The war porn video's visual climax includes one of the most explicit scenes of an enemy fighter's death. It shows (at 2:50–2:59) an armed dark-clad insurgent, set against a light-colored stone wall, and engaged in intense street combat. The establishing shot has him violent and effective—in a crouching warrior position, getting off several rounds with an AK-47. After firing to his left, the fighter stands, pauses, and looks away to his right, upon which he is obliterated by a rocket or other large explosive. Several "Let the Bodies Hit the Floor" comments seized on the spectacular nature of the fighter's death. "2:55 best scene!!," said TheCarl1001 in 2011. "The terrorist that gets blown into pieces is far fucking better than any bullshit Hollywood movie," turkzil added in 2011. "This is one reason," the viewer continued, "why I want to join the marines."[57] And this is an apotheosis of the war porn genre. Here is a worthy and dangerous target. At the viewer's street level, one dimly perceives his manner, carriage, and facial expression as he rests from firing. And then he is obliterated, with no trace left behind.

Pornographic enthusiasm for killing the unknown enemy also arises in the ever-popular videotaped gun-sight footage of pilots zeroing in on and taking out targets from the air. The Nintendo-like image that U.S. bomber crews saw on their screens in the Gulf War and that was then broadcast by CNN and other television networks had by the Iraq War become utterly ubiquitous but still enthralling, as the images so seamlessly merged wartime killing with entertainment value, and combat with spectatorship. Considerable YouTube war porn integrates these aerial shots into videos largely focused on ground combat. The aforementioned "Iraq: Bird Is the Word" video transitions from a scene of marines patrolling a city street to the grainy moving picture from an aerial gun sight (at 0:57–1:06), its black and white negative imagery identifying two humans on the ground, both of whom are shot and killed by machine gun fire. In contrast to the killing in "Let the Bodies Hit the Floor," these targets are visually indistinct, noticeably human only by their vague bodily outline and movement. There's an ethereal, orbish quality to their white radiance, though.

And their elimination in a hail of bullets excites many YouTube viewers: "OHHH MYY I LOOOOOLLD SO HARD AT 0:57 t0 1:05"; "1:00 OWNED!"; "0:56 YEA MOTHERFUCKAAAAAAA THAKE THAT IN DAH FACEEE!!!!"[58]

This practice among YouTube commenters of pinpointing certain scenes and moments of enjoyment helps the contemporary historian estimate the impact of war porn on its viewers. Like YouTube war porn discourse in general, time-stamped comments vary in content and tone: from the dramatic "NOOOOOOO 0:33 FUCK US !!!!!!!!!!!!!!!" typed in protest of a marine shooting and killing an unarmed and wounded insurgent fighter, to incidental observations about overlaps in military visual culture. An "Iraq: Bird Is the Word" viewer, for instance, pausing on the aerial targeting of Iraqis discussed earlier, recognized war porn's intertextuality with combat video games: "0:57 Reminds me of modern warfare 1."[59] Some comments focus on the countenance of marines in action. A viewer of "Bullet with a Name on It" was struck by the intensity of a marine on patrol in an Iraqi city street: "Pause it at exactly 1:37 and look at that US Warrior holding his SAW Machine gun, on the hunt for combant [sic]. Kinda freaky. Hooah!" Another "Bullet with a Name on It" viewer noticed the happy absorption of a marine in combat: "0:42, I love the guys smile! It is like sayin 'I love my job' ;-)."[60] Such comments are telling in that they provide insight as to how war porn is actually taken in on YouTube and other video-sharing sites. By clicking pause, viewers savor and reflect on particular images, which are often replayed for added enjoyment and meaning. The pinpointed comments on individual marines' expressions also demonstrate the deep interest in the subjective experience and sensibility of war. In a familiar tenet of cultural studies, these viewers tend to find what they are looking for: in addition to destruction, gore, and high-tech weaponry, they see that U.S. troops fight with resolve, purpose, and pleasure.

As to pleasure in war, YouTube war porn offers it in spades, as seen in the videos' subjects in action and documented by the posted comments upon those images. Some postings require the historian to intuit a viewer's satisfaction. Among the responses to "U.S Marine Kills Iraqi," for instance, a viewer in 2012 wrote, "That's what happens when you send US marines to fight, Killing is their job and we expect nothing less than that." Another 2012 viewer of "Iraq: Bird Is the Word" commented: "Great vid. Nothing better than to watch some dune coon muslim terrorists blown the fawk up. You can't get any better than this."[61] Unabashed pride in Marine violence sits just beneath the surface. Other comments are utterly transparent, almost perfectly complete in registering a viewer's affect. "This video gave me the chills," a viewer of "Bullet with a Name on It" wrote in 2009: "US marines really kick ass :D." The most emotive comments usually do go to strong feelings for the Corps itself. In 2006 another viewer of "Bullet with a Name on It" paid the video perhaps the ultimate compliment: "Oooooofuckinrahhhhh!!! Makes my dick hard each time I watch this video. Semper Fi."[62] Whether or not the viewer was speaking literally, the pornographic effect of Iraq War combat videos seems clear.

The Iraq War on Television

In comparison to that of World War II and Vietnam War combat films, the violent pornographic payoff of YouTube videos from Iraq is more immediate, less contextualized by story line and character development. Iraq war porn is without certain conventions of classic Hollywood cinema that add to feature-length combat films' pornographic effect. Iraq war porn is without, for instance, the close ups of actors' faces in battle scenes that vivify and model the pleasures and other subjective registers of war. (YouTube comments on marines' facial expressions seen in a few war porn videos come from viewers who, having undoubtedly watched so many hours of Hollywood combat films, concentrated on those shots that approximated what they had cinematically come to expect from moving pictures of war.) In contrast to the high production value of Hollywood combat films, YouTube Iraq War videos' strong suit is contemporaneity, an authenticity that comes from being sourced amid the very conflict they represent.

Sitting between the artifice of Hollywood combat movies and the immediacy of YouTube war porn is the television programming that focused on the Iraq War, a corpus of material ranging from network news coverage to prime-time dramatic series. A common denominator of U.S. TV's treatment of the Iraq War is "militainment," as media critic Roger Stahl puts it: "war packaged for pleasurable consumption," through which the attentive citizen turns into pliant viewer, exchanging critical engagement for surface drama and spectacle, and thereby making way undiscerningly for American empire.[63] Recognizing the entertainment value and hegemonic function so central to Iraq War programming is not to say, however, that all of the TV shows mindlessly celebrated American firepower or even forwarded the Bush administration's *casus belli* of weapons of mass destruction or freeing the Iraqi people. Iraqi oil reserves and an overarching will to global economic power are contemplated by marines and soldiers in the two series on combat in the war, *Over There* (2005) and *Generation Kill* (2008). Moreover, both shows take up the tragedy of civilian casualties caused by excessive and indiscriminate U.S. force. Even these series' critical engagement goes only so far, though. Anguish over civilian deaths is as much about losing military advantage as it is over humanitarian concern. *Over There* and *Generation Kill* tend to preclude large issues like just war or distribution of physical risk and harm, as Stacy Takacs points out, and instead forward "a morality of action"—a fairly bloodless sensibility of military professionalism and expertise that still turns on capacity and motivation to kill.[64] "It's killin time," as the levelheaded Sgt. Scream tells his Army platoon in *Over There*: "Somebody's going to die today, and that's a fact. The only question is who."[65]

Iraq War programming intersected with one of the largest "taste fractions" in the multichannel satellite and cable conglomeration that made up early twenty-first-century commercial television.[66] With sports coverage long a mainstay of network television, male-directed programming begat the hypermasculinist niche audience of post–September 11 America, where monster truck racing, mixed martial arts, and alligator wrestling could be viewed along with live-feed updates on Fox News of the Iraq invasion, for instance or, later, an A&E documentary on the football star turned Army Ranger Pat Tillman. During this time, the Department of Defense would found and run three cable channels—the Pentagon Channel, the Military Channel, and the Military History Channel. No better gauge existed for the militarization of American culture and society than the programming on commercial TV. In celebrating U.S. military service, Fox's *War Stories with Oliver North* (2001–10) went "behind the headlines and beyond the history books," as the show's website describes, "to hear veterans recount in their own words the missions, battles, and intrigue from America's conflicts."[67] Spike TV—with shows like *MANswers: The Ultimate Male Survival Guide* (2007–11), a one-station pantheon to American hypermasculinity—aired *Deadliest Warrior* (2009–12), which analyzed through dramatic "reenactments" the prowess of well-known armies in history; episodes such as Viking versus Samurai and Nazi Waffen SS versus Viet Cong evaluated each force's weaponry, tactics, skill, and will to kill in battle and then—through blow-by-blow simulation—ventured the likely winner in a fantastically counterfactual confrontation between a fighter from each side.[68]

While *Deadliest Warrior* somehow overlooked the Marine rifleman, the U.S. Marine Corps was otherwise paramount in Iraq War–era military programming. From the Discovery Channel's documentary *Making Marines* (2002), to

Oliver North's several episodes on marines in the Iraq invasion, to Showtime's early postwar drama *Homeland* (2011–), television during the Iraq War certainly extended the Marines' reputation as the embodiment of America's warrior class. Heading this show of prominence was *Generation Kill*, a detailed and complex study of Marine firepower in Iraq and the ethos that went with it. The HBO series closely followed *Rolling Stone* reporter Evan Wright's memoir, whose thesis is stated in the book's full title: *Generation Kill: Devil Dogs, Iceman, Captain America and the New Face of American War*. There's a cultural coherence to the young Americans invading Iraq, Wright argues: "These Marines would be virtually unrecognizable to their forebears in the 'Greatest Generation.' They are kids raised on hip-hop, Marilyn Manson and Jerry Springer." They "represent," as Wright expands his sociological profile, "what is more or less America's first generation of disposable children." With "more than half of the guys com[ing] from broken homes . . . many are on more intimate terms with video games, reality TV shows and Internet porn than they are with their own parents."[69] As the title claims, what most unites them generationally is the ability to kill. Their platoon commander, Lieutenant Nathaniel Fick, notices it right way. The Dartmouth-educated officer implicitly compares them to the subjects of S. L. A. Marshall's study:

> I'll say one thing about these guys. . . . When we take fire, not one of them hesitates to shoot back. In World War Two, when Marines hit the beaches, a surprisingly high percentage of them didn't fire their weapons, even when faced with direct enemy contact. They hesitated. Not these guys. Did you see what they did to that town? They fucking destroyed it. These guys have no problem with killing.[70]

More so than the book, the HBO series illustrates this violent fact of generation kill. Produced by *The Wire*'s David Simon and Ed Burns, the TV series locates the impetus of the Marines' destructive force at an institutional level, with commanding officers ramping up the pace and scale of attack to impress the generals with their aggressiveness. "The violence of action is to our advantage," Battalion commander, Lieutenant Colonel Stephen Ferrando, tells his company and platoon leaders in episode 3, "Screwby." Meanwhile, the Marine grunts embrace the imperative of action, with some reveling in the opportunity; as Lance Corporal Harold James Trombley says in the same episode, with a big smile on his face: "I'm shooting motherfuckers like it's cool."[71] It's not clear if those who physically survive the fighting will otherwise make it home in one piece. Glimpses of anguish and trauma come through amid the carnage. But, finally, as *Generation Kill* focuses only on the invasion, attention to cause and effect regarding the mental and moral casualties of war is minimal. Instead, and again, *Generation Kill* pursues a "morality of action." It fixes on (what the medium does best) the imagery of modern warfare, as in episode 5's ten-minute scene of artillery exchange and firefight, all seen by way of the marines' night vision goggles. "Cool" indeed: these spectacular shots—images that make the corresponding scenes of, say, *Apocalypse Now* seem almost flat— can become something of a world unto themselves, with the depicted marines' excitement and wonder transferred to the viewing audience.

A show that aired two years before *Generation Kill* does take in a fuller picture of the impact of the Iraq War on one marine at least, Sergeant Matthew Orth. In "For God and Country: A Marine Sniper's Story" (2006)—an hour-long profile aired on MSNBC, produced by the NBC's news division, and narrated by the network journalist and weekend anchor Lester Holt—Orth testifies to the heavy burden he carries after having killed "hundreds" of Afghanis and Iraqis. There is a poignant quality to Orth's story. The super-close-up interview shots have this young man worn and troubled. When he says, "People don't understand what we go through over there," one is inclined to believe him. Orth describes the positive arc of his life, from joining the Marines because he thought "war was cool" to having "a lot of respect for life now" because of the violence he saw and wrought. He "did a lot of growing up" while in war. Upon his return home, Orth reestablished a relationship with his father, something that had been missing since his childhood. He states that when he has children, "I want to raise them to know that there are other cultures out there and to value life."[72] In Orth we meet a Marine sniper who is proud of his service—he had the longest confirmed kill in the conflict; and, at the same time, he is somewhat repentant if not ashamed of the damage and death that he caused.

"For God and Country" prompts several responses. It is possible, on the one hand, to respect the subject and care about his humanity, while still shuddering over the violence he wielded, and, on the other hand, to find the MSNBC broadcast cynical and, finally, pornographic in its intent and message. In reintegrating into civilian life, Orth describes, "the hardest thing was to be a loving individual." The comment has weight, gives pause. And, yet, Orth's distinctive role as a killer, as a sniper "assassin," is exactly the premise on which the show is based. Few people in this world are actually allowed to kill in cold blood, "Marine Sniper" essentially says; we found one; and we're going to use him to tell you, as Holt promises, "what it feels like." In the network's introduction of

Orth, using exaggerated graphics of a sniper rifle sight targeting enemy fighters, there's nothing modest, critical, or repentant. It's most interested in the sniper's hardness and the finality of his actions. "One shot, one kill," Orth is quoted as saying in the show's opening, with the words then superimposed white on dark gray. "He must be prepared to take a life," Holt narrates gravely, "no matter what, because *that* is his job." The show then cuts to a grisly scene of a dead mujahedeen fighter in the middle of a city street. There's a "it's a tough job but somebody's got to do it" sensibility to MSNBC's accounting. Orth is world weary. His guilt only adds to the gravity of his task. The point is underlined by Orth's repeated observation, "We really do play the hand of god." Or, as Holt, melodramatically declares: "His call sign, the Grim Reaper." Toward the end of the hour, Orth recalls his reticence over telling his friends about all the killing in combat. He "was very surprised," though, by "how they reacted." They were "hungry" for it, he continues: "I had all these bloody pictures and videos of dead Iraqis. They wanted to see everything." NBC knew better. The network banked on its viewers' appetite for war.

A leading supplier of war on television during the early twenty-first century has been the History Channel. Founded in 1995, the satellite and cable TV channel made its name and built its audience through military programming. Critics dubbed it "The Hitler Channel" because of its fixation on World War II. Following September 11, the History Channel filled its schedule with war programs, led by "Fighting Fridays," a primetime lineup of its most popular in-house produced shows. *Lost Evidence* (2004), for instance, uses three-dimensional graphics, aerial photographs, and oral histories to re-create key battles of World War II. Similarly, *Dogfights* (2006–7) plays out battles between fighter pilots through real footage, interviews, and digital simulation. In a different vein, *Mail Call* (2002–9) features R. Lee Ermey, former Marine sergeant, and the actor who played drill instructor Sergeant Hartman in *Full Metal Jacket*; in the TV show, Ermey answers viewers' questions regarding weapons, military training, and the like, while playfully deriding the letter writers and generally staying in DI character. In a more recent addition to Fighting Fridays, the History Channel aired *Full Metal Jousting* (2012), an actual jousting tournament between sixteen contestants. Amid the History Channel's Friday night martial lineup, *Shootout* (2005–6) stands out as one of its most talked about shows, and as one that depicted particular battles in the Iraq War.

The Iraq *Shootout* productions are notable for their contemporaneity. The first episode, "D-Day Fallujah," aired just a few months after that November

2004 battle, close on the heels of the amateur-produced combat videos from Fallujah placed on the internet. Adding to the immediacy of the History Channel account was the close cooperation of the Marine Corps and of Third Battalion, First Regiment—a key unit in the Fallujah battle, and the same one featured, as discussed earlier, in one of the most viewed YouTube videos of the Iraq War. This first *Shootout* episode focuses on the 3/1's mission of clearing the Jolan district of Fallujah. Marines from the unit, just back from the battle, are interviewed for the show, and particular firefights in the operation are reenacted in blow-by-blow fashion based on their stories.[73] Having established broad geographic and strategic context with maps, photos, and interviews, the series' signature production technique is to depict the physical space of the combat with three-dimensional graphics of a certain house or city block and then reenact realistically the most crucial moments of the fighting, intercut with further diagrams charting the combatants' changing positions. As one viewer commented in 2012 on YouTube, "Great documentary!"[74] And "D-Day Fallujah" does achieve a solid documentary effect. It establishes a sense of actuality, pulling the viewer ever closer into the fighting, with the 3/1 veterans attesting "this is how it was." Adding to its realism is the interspersion of video material taken by marines in Fallujah with street shots from a set in the United States, replete with costumed marines, Iraqi civilians, and insurgents. It's sometimes hard to distinguish between the actual and staged footage.

Like most documentaries, "D-Day Fallujah" has a subjective investment, a promotional push. But just what is it promoting? The U.S. Marines come across rather well. This is hardly surprising, given the Corps' cooperation with the production and the History Channel's uncritical bearing. The grunts deserve our respect and veneration, the show assumes. In its summing-up segment, a Marine lieutenant colonel insists, contrary to Evan Wright, "There's no difference in this generation." His Marines "performed as you read about any generation performing," he continued, referring back to that baseline of heroic service, the generation of World War II marines and soldiers: "They're still great." But with few doubting the Marines' martial prowess, "D-Day Fallujah" doesn't devote too much to it, at least at first.

What the show zeroes in on and sells is the ferocity of the combat in Fallujah, and for this it must establish the *insurgents'* deadliness as a fighting force. "Most of the insurgents," the voice-over narration explains, "are foreign fighters. And," it continues after a dramatic pause, "they're here to die. So they have no plans to run when the Marines come knocking. They've come to Fallujah to martyr themselves, and to take as many Americans with them as possible." Cut to the 3/1's Lieutenant Jesse Grapes, who confirms that "these are guys who are experts at their profession, they were true warriors. . . . They knew what they were doing. They knew how to use weapons. They understood the principles of an ambush. Really good fighters." With shots of actors dressed as insurgents, handling and firing their AK-47s and other weapons in mock city streets, the narration adds that they're especially dangerous. "As well trying to defeat the U.S. troops' excellent body armor, the insurgents take aim directly at their faces." As if that wasn't enough, many had prepared for battle by "getting high on narcotics, like liquid adrenaline, which gives them an almost superhuman capacity to withstand pain" and, as we'll see, to continue fighting after being mortally wounded.

Having established the worthiness of the Marines' enemy, much like sportscasters build up the prowess of both sides before a big game, "D-Day Fallujah" proceeds to illustrate the much anticipated and vaunted urban combat in Fallujah. "It is the deadliest house to house street brawl since Hue City Vietnam," as the show sells itself, "close quarter combat of the scariest kind." If the show can be called a documentary, its subject is really quite narrow: what it's like to clear houses of heavily armed hopped-up insurgents. It works from a single point of view—the Marine rifleman. So, while emphasizing Fallujah's dangerousness, we get that from just one perspective. As the Marines took casualties, they started to rely more on the firepower of explosive charges and Abrams tanks. They called in bulldozers to level buildings housing insurgents. To have a house crash in on you from all sides would make for exciting TV, but the show isn't interested, as it has the insurgents' fate already determined. What matters is the accomplishment of that predetermined outcome. And here the show excels. Its obviously staged firefight scenes are, nevertheless, vivid, fluid, and detailed. While there's some gore in the form of bloody dead insurgents, the pornographic punch is one of simulation—that is, of firing an M-16 at heavily armed targets, most hidden, some moving, and all of them shooting back at you. The point of view shot looking down the barrel of an M-16 rifle, as it moves to target insurgents, is taken directly from first-person shooter combat video games. As one YouTube viewer of "D-Day Fallujah" wrote in 2012: "Damn, this makes me want to play COD [the videogame *Call of Duty*]."[75]

If History Channel executives harbored any pornographic intent in airing "D-Day Fallujah"—that is, in prompting its viewers to want to engage in actual fighting—then they probably wouldn't have been too disappointed with the commenter's urge to play a combat video game. As examined in the following chapter, the two behaviors of video gaming and warfighting became very closely entwined in the early twenty-first century: while the former helped recruit and train marines and soldiers for the latter, the latter directly informed the content of the former. Between 2003 and 2006, riflemen of the Third Battalion, First Regiment—the same battle-hardened unit featured in the HBO show on Fallujah—were ordered by the Marine Corps to work with a video game producer hired to make two different titles focusing on urban combat in the Middle East. The loop between cultural text and conduct of battle would grow ever tighter in combat video games, war porn videos, and television programming. "USA has always been good at two things," as a 2006 YouTube viewer of the "Bullet with a Name on It" video perfectly stated the point: "Making war and entertainment. There doesn't seem to be much difference between these two things anymore and both feed each other."[76]

CHAPTER THREE

Fallujah, *First to Fight*, and Ludology

Fallujah, Al Anbar Province, makes up a centerpiece of both the Iraq War and this study of the U.S. Marines' force projection in Iraq through particular optics of combat and overarching regimes of seeing. Within the visual culture of the post–September 11 wars, for instance, few images have been more startling to American eyes than those from March 2004 of the charred and mutilated bodies of four Blackwater contractors hung from a Fallujah bridge. Added to that, the United States halted the first Fallujah battle in April 2004 out of concerns about the stories and images of Iraqi civilian casualties. The NBC videotape of a marine shooting an unarmed and wounded insurgent in a Fallujah mosque during the November 2004 battle became a symbol of excessive U.S. military force, as TV channels, YouTube, and other media outlets replayed it throughout the world.[1] The *Los Angeles Times* photograph from the Fallujah battle of "the Marlboro Marine"—Lance Corporal James Blake Miller of the First Battalion, Eighth Regiment—became an immediate icon of the toll that combat takes on its survivors. Closest to this chapter's focus, the controversy over the combat video game *Six Days in Fallujah*, based on the Marines' fighting in November 2004, highlights the centrality of video gaming and simulation in the Corps' early twenty-first-century culture.[2] While *Six Days in Fallujah*'s publisher decided to drop the title amid outcries from family members of marines and soldiers who fought and died in the city, the game's production marks both the Corps' long-term investment in computer simulation for training and the collaboration between Marines and commercial video game makers.

The same video game studio that made *Six Days in Fallujah* had worked closely with the Marines in producing *Close Combat: First to Fight* (2005)—a first-person shooter that, per the Corps' specifications, centered on the challenges faced by a four-man fire team in urban combat. For both *Six Days in Fallujah* and *First to Fight*, video game designers engaged marines from the Third Battalion, First Regiment between deployments in interviews about small-unit fighting in Iraq's city streets.[3] Destineer Corporation would market *First to Fight* to recreational players on its authentic battle experience—the gold standard of combat video games—while the Marines received from the partnership "tactical decision making simulation" for "small unit leaders." In the language of its Modeling and Simulation Management Office, the training system "will emulate the tactical combat environment and allow squad leaders, team leaders, and team members to practice the appropriate cognitive skills in a first-person synthetic environment." For the Marines, the most important cognitive skills involved fighting "through streets under siege"—the perception, movement, and decision making for "tak[ing] down rooms, achieving multiple angles of fire against enemies," and keeping "their fingers firing their weapons" while "bring[ing] their teams safely through battle."[4] In working with Destineer, and in assigning 3/1 riflemen to basically "upload" their combat experiences into the game, the U.S. Marine Corps created a system that would hone warfighting skills in recreational gamers and marines alike.

First to Fight would be released commercially and distributed to the Marines in spring 2005, just months after the battle for Fallujah and amid the occupation and ongoing fight for control of the Sunni city.[5] What's more, an uncanny congruence can be seen between the content of *First to Fight* play—its focus on clearing buildings, for instance, and its ready use of air support—and the Marines' conduct of battle in Fallujah in November 2004. As we continue to examine the ways in which visual culture constituted the U.S. Marines' making of war in Iraq, the combat video game *First to Fight* presents an integral example of the tightening loop between visual representation of battle and actual fighting and killing. After the battalion's second deployment to Iraq in 2004–5, which included a central role in the Fallujah battle, 3/1 marines approached the same video game producers, urging them to make a sequel that would document the hardship and heroism of what would become an epic moment in Corps history. After considerable capital investment and creative effort went into producing the third-person shooter, *Six Days in Fallujah*, the Konami Corporation decided not to publish it after parents of marines and soldiers killed there charged that "they are trivializing the battle" by "making it something people play for fun."[6]

Ironically, as this chapter argues, a keen sense of play contributed to the Marines' effectiveness in Fallujah. Play and war are not mutually exclusive. "Ever since words existed for fighting and playing," the Dutch historian and anthropologist Johann Huizinga commented famously in his book *Homo Ludens* (1938), "men have been wont to call war a game." For Huizinga, play makes up a primary element of human nature: closely allied with "agon"—competition against an equally matched opponent—play helped shape war into the cross-cultural human institution that it is today. War is "the most intense, the most energetic form of play," Huizinga stated, "and at the same time the most palpable and primitive." Having seen the ravages of World War I, however, and writing on the eve of its successor, Huizinga believed that the "play element" in war would soon be "extinguished" by total war's mass slaughter and industrial composition.[7] He got that last part wrong. In one of modernity's great tragedies, the play element in war lived well past world war. It's alive and well today. Not that it permeates every aspect of military doctrine, tactics, and warfighting. But, along with sport, play remains a tremendous lure to war.[8] It animates preparation for battle. And while most evidence suggests that playfulness dissolves once fighting begins, the notion of combat as a contest between equals continues to resonate in the vast cultural representation of war. It's at the center of warrior culture. And it certainly could be found among the Marines in Fallujah.

The November 2004 battle for Fallujah was just that, a set piece battle in which U.S. Marine, U.S. Army, and Iraqi ground forces fought to clear some two thousand insurgents and foreign jihadists who were determined either to hold their positions or to die trying.[9] In preparing for the assault, Fallujah was constantly likened to Hue City, the 1968 Vietnam War battle in which three Marine battalions fought North Vietnamese regulars and Viet Cong in brutal house-to-house combat for more than twenty days. Hue City had become ingrained in Corps legend, and the Marines expected Fallujah to follow. "This is one of our largest fights since Vietnam," Colonel Michael Shupp, commanding officer of Regimental Combat Team 1, told his marines before attacking. "As time goes by," he continued, "when we celebrate the Marine Corps birthday, Fallujah will be just as significant as the battle for Inchon, the Chosin Reservoir, Khe Sanh, or Hue City. It will ring in Marine Corps history as another great epic battle we fought in war," Shupp concluded: "Do your worst. Kill everything in front of you."[10] In 2006 Al Anbar Province would see the "Awakening," the Sunni turn against Al Qaeda Iraq that the Marines helped along via counterinsurgency efforts of economic reconstruction, political cooperation, and cultural awareness; Fallujah in November 2004, as Shupp's speech suggests, followed only the first violent tenet of counterinsurgency doctrine: kill the "bad guys," the armed fighters set among civilian population. And with Fallujah's residents told to leave days and weeks before the attack, the Marines brought its full measure of force to bear upon those remaining in the city, which still caused approximately eight hundred Iraqi civilian casualties.[11]

The Marines' projection of force in Fallujah started with air power. The Third Marine Aircraft Wing and Air Force formed a "wedding cake" over the city: "layered all the way up," low-flying helicopters, fixed-wing attack jets, AC-130 Gunships, and unmanned aerial vehicles (UAVs) poured cannon fire, rockets, and bombs on targets identified by forward air controller teams deployed with infantry units below.[12] Target acquisition also worked through the forward-looking infrared (FLIR) and daytime optical cameras mounted on UAVs flying at several thousand feet; the Marine Unmanned Aerial Vehicle Squadron 1 operating ten miles outside of the city employed line-of-sight video feed to coordinate aircraft and artillery fire against enemy positions.[13] Aircraft and artillery used white phosphorous munitions. Two-thousand-pound bombs dropped on Fallujah's railroad tracks triggered the ground attack. Abrams tanks, Amphibious Assault Vehicles, and Bradley fighting vehicles provided heavy firepower with cannons, grenade launchers, and .50 caliber machine guns. Added firepower came from "assault teams" attached to infantry platoons. Armed with shoulder-launched multipurpose assault weapons, Bangalore torpedoes, and satchel charges, assault teams blew holes in buildings hiding enemy fighters and destroyed weapons caches. The wholesale leveling of enemy positions came to be a preferred Marine tactic, one best accomplished with Armored Caterpillar D9 Bulldozers and M9 Armored Combat Earthmovers.[14]

The main U.S. force in the battle consisted of two Army battalions and four Marine battalions arrayed along Fallujah's northern border and ordered to move south while clearing insurgents and jihadists from their positions in the city's houses, mosques, factories, and streets. It *was* a major operation—"the fiercest battle inside houses," as Bing West has described, "that I can find in recorded history."[15] Above all else, Fallujah meant urban combat: warfare in contained space amid civilian populations that delimited American forces' superiority in numbers and firepower; but, at the same time, it epitomized a type of fighting that the Marines had been anticipating and training for since the mid-1990s. In an institutional sense, the Marines had long been looking forward to Fallujah. To be sure, some within the I Marine Expeditionary Force weren't crazy about returning to Iraq and Al Anbar in 2004 after MEF's breakneck march up to Baghdad the year before: occupying a province the size of North Carolina with thirteen cities, two million civilians, and a growing insurgency isn't what assault troops do best.[16] But with the March 2004 Blackwater killings in Fallujah, the Marines' counterinsurgency in that city took on a decidedly aggressive posture. After being ordered to halt a first attack in April, the Marines had become all the more eager to complete the November mission against the well-armed and entrenched insurgents. Fallujah posed both a challenge and an opportunity for all-out, close-quarter combat in an urban environment.

Marines like a good fight, as the saying goes: they "pray for war." And in assessing the force projected in Fallujah, that ludic tenet of wanting to prove themselves against a worthy enemy needs to be included. It's, in part, about play and competition and the drama of war. Third Battalion, First Regiment marines reportedly felt chills as they moved from their staging position just north of the city and, on November 8, 2004, crossed the line of departure with Psych Ops blasting Wagner's "Ride of the Valkyries" from its Humvee megaspeakers.[17] If we go back to the days preceding the battle, back to Camp Fallujah, Camp Baharia, and Camp Abu Ghraib, where the marines awaited orders to attack, the historical record of the marines preparing for the Fallujah battle is chock-full of exuberance and enthusiasm for the fighting ahead. "This is what people join the Marine Corps to do," to repeat the words in a YouTube video of a 3/1 marine looking forward to the battle for Fallujah: "You might be in the Marine Corps for twenty years and never get this chance again—to take down a full fledged city full of insurgents. . . . If you want some, get some."[18] Two days before the battle started, Lieutenant Colonel Willard Buhl, 3/1's commanding officer, staged in Camp Fallujah a "First Annual 'Ben Hur' Memorial Chariot Race," replete with confiscated Iraqi horses and cobbled-together carts.[19] In what is perhaps a more pedestrian example, we know that marines had their Xbox consoles, Gameboys, and laptops in camp to play *Doom* (1993) in its various modifications, *Wolfenstein* (2003), and other combat video games.[20] If asked why they were playing, some marines would have undoubtedly said it's to relieve the tension or that it's something to do, while others might have answered simply that it's fun.

Ender's Game and the Rise of Simulation in Military Training, 1995–2005

Within the social scientific and humanistic study of video gaming, some tend to see the games as creative expression like books and films, while others approach them as activity and play like sports and board games. "Narratology v. ludology" is one label for the academic split.[21] Whether one understands a combat video game to be something like, on the one hand, *The Iliad* and *Full*

Metal Jacket, or, on the other, playing football and chess matters because the difference between the two conceptions informs the critical question of whether playing violent video games leads to violent behavior. In 2010 the U.S. Supreme Court weighed in on the issue in *Brown v. Entertainment Merchants Association*, with the majority decision treating video games as closely akin to literature in ruling a California statute restricting their sale to minors unconstitutional as an abridgment of freedom of speech.[22] If the Court had instead likened video games to play, it may very well have upheld the California law. The First Amendment doesn't protect play like it does speech. And to view video gaming as play rather than speech would have privileged an increasing body of research that shows that the distinct nature of interactivity in the medium does indeed stir hostile and aggressive urges. Of course, literature and film are also interactive. As argued in the previous chapter, Hollywood combat films, YouTube war porn, and cable television shows depicting fighting in Iraq can create desire in the viewer to engage in similar behavior. Something more, however, is involved in video gaming. Or so the ludologists contend.[23]

The Marine Corps would seem to be among those who disagree with the Supreme Court's narratologist view of video gaming. Nobody calls the Marines ludologists. Given the millions of dollars it has invested in interactive video systems, the Corps itself prefers the seriousness of the word *simulation* over *games*. Yet play or ludology has been central to the Marines' approach to warfighting. More broadly, the Department of Defense is clearly committed to video gaming's effectiveness in preparing recruits for combat. In a dissenting opinion in *Brown v. Entertainment Merchants*, Justice Stephen Breyer stressed that, compared to books and films, video games are extremely adept tools for inculcating the performance of new tasks. Breyer used the military as his lead example: "Why else would the Armed Forces incorporate video games into its training?"[24] That is exactly the point. While lawyers, preadolescent gamers, and professional psychologists challenge the causal relation between playing video games and violent behavior, the Pentagon has banked on it, developing since the early-1990s close partnerships with universities, corporate video-game producers, and the digital entertainment industry overall.[25] And behind this enterprise stands considerable intellectual investment in the idea that human behavior can be shaped by play within increasingly realistic replicas of the material world and the experience that it entails.

The vision driving the military's video-game development has largely been one of science fiction: to unify the real world with simulation. "Virtual reality research," as the technical director for the Army's simulation program Michael Macedonia commented in 2004, has put the U.S. military in constant quest for closer and closer "copies of the real world." Science fiction "has had a lot of influence on our thinking," Macedonia has acknowledged, pointing first and foremost to Orson Scott Card's book *Ender's Game* (1985; serialized in 1977), a foundational text during the turn-of-the-century integration of civilian entertainment with military training. To consider the book's plot is to imagine simulation's capacity for creating killers. As Macedonia has said suggestively: "I've always been fascinated with what you can do with a six year old."[26]

Macedonia's reference is to the child Andrew "Ender" Wiggin, the title character in Card's futuristic story of defending Earth from an anticipated attack by the Buggers, an alien race whose superior numbers nearly destroyed humankind in two previous wars. The International Fleet recruits Ender for its Battle School, where his genius for tactics and leadership speed him through training and cast him as the "only hope in the upcoming invasion." Promoted to Command School, Ender continues his training on the "master simulator." Amid a culture saturated by "vids," the master simulator is "the most perfect videogame he had ever played." Ender also takes classes in traditional subjects, "but the real education was the game"—an elaborate holographic star wars display that is constantly upgraded to "more closely duplicate the conditions you might encounter in a real battle, where you will know only what your ships can see."[27] The boy's education culminates under Mazer Rackham, the aged mastermind of the earlier victories over the Buggers who tells Ender that he will fill the role of the enemy for his student's last and most challenging war games.

Ender's final examination puts him in command of a small fleet of ships headed toward the Bugger planet protected by a force outnumbering him a thousand to one. Desperate and dismayed over the test's unfairness, Ender "cheats" by firing a nuclear weapon that would annihilate the planet and all of its inhabitants. Believing that he has flunked, Ender quickly learns of the teachers' utter joy over his decision and the ensuing victory. "You made the hard choice, boy," Mazer says in praise: "All or nothing. End them or us." Still confused, Ender insists: "I beat *you*." Mazer laughs heartily and explains that Ender never played him: "There were no games, the battles were real, and the only enemy you fought was the Buggers."[28] In the final days the master simulator had been no simulator at all but rather a full-screen representation of the enemy and, with that, the medium for the Buggers' destruction.

Ender's Game recognizes that simulation minimizes moral responsibility for killing as well as hesitation over the act itself. "You tricked me into it!" Ender screams at Mazer and his other Command School teachers. "Of course we tricked you into it," one responds: "That's the whole point. It had to be a trick or you wouldn't have done it." But "I didn't want to kill them all!" Ender insists: "I didn't want to kill anybody! I'm not a killer!" Mazer agrees with him: "You had to be a weapon, Ender. Like a gun, like the Little Doctor [the nuclear weapon], functioning perfectly, but not knowing what you were aimed at. We aimed you. We're responsible. If there was something wrong, we did it." Despite his unusual capacity for killing—and the fact that his actions obliterated a whole species—Ender does maintain some semblance of innocence in others' eyes if not his own. "Any decent person who knows what warfare is can never go into battle with a whole heart," Mazer continues: "But you didn't know. We made sure you didn't know."[29]

Along with tricking Ender into killing, the novel hinges on the truth that the older generation responsible for taking a people to war depends upon its youth to achieve that conflict's violence. *Ender's Game* fixes on the younger generation's superior vision and reflexes—attributes that have long replaced physical strength and ferocity as crucial to martial success. "It had to be a child Ender," Mazer says while trying to justify his deception; "you were faster than me. Better than me." Like a latter-day pinball wizard, only Ender could have pulled it off. Having spent the better part of his short life playing the most current and challenging video games, the conditioning of Ender's young neurological pathways had cultivated deadliness far beyond his elders. As his teachers well knew, Ender would always be drawn to the video screen. Despite or rather as part of his keen intelligence, Ender "lives for" and finds continual "pleasure" in video games. Key words for Card, the *pleasure* and *play* of combat—"It was pleasure; it was play"—provide the means for replicating fighters. "Battles became exhilarating" for Ender, as they had for Mazer before him. Spurred by the challenge it entails, Ender discovers in battle the age-old, intergenerational bond through war. "You were reckless and brilliant and young," Mazer praises Ender for his student's annihilation of the Buggers: "It's what you were born for."[30]

Writing *Ender's Game* at a nadir of American militarism, shortly after the Vietnam War, Card realized that the pleasure principle in combat could still be tapped in the nation's youth, while at the same time he predicted that the state would rely on technological innovation to make killing in future wars more remote and less destructive for the killer. Card was prescient. In imagining the course and implications of that enterprise, he had only a few decades of very basic digitalized military simulation from which to draw, a short history that began with collaboration between the U.S. Navy and the Massachusetts Institute of Technology in the 1940s to combine computer, control console, and movie screen for training pilots. Simulation would develop slowly during the Cold War. The Pentagon focused on conventional and nuclear war primarily for deterrence and for fighting hot wars in Asia and elsewhere. By the 1960s the entertainment industry had started to build war games for the multimillion-dollars-a-year arcade business. A prototype for the turn-of-the-century first-person shooters is the *Midway Sea Raider* (1969), an electronic World War II submarine game in which the player uses a periscope surrounded by a sonar-beeping console to fire torpedoes at different-sized boats moving horizontally across the screen. One has to estimate the different speeds of the cruisers, destroyers, and battleships and then "lead" the target to score a strike. As the manufacturer's print advertisement describes the resultant sense impression, "A hit becomes a vibrating explosion of sound and color."[31]

As *Space Invaders* (1979) and *Pac Man* (1980) drew more and more young video-game players into the arcades, Atari's *Battlezone* (1980) put the fighter's-eye view behind a large gun turret and added depth of vision and full range of movement through three-dimensional space (fig. 3.1) With simply-drawn mountains, a smoldering-volcano, and a crescent moon set in the background, solid cubes and pyramids dot the landscape, a desert-like plane, where one maneuvers through the geometric obstacles and shoots at enemy tanks with two joysticks and a button trigger. *Battlezone*'s depth perception and first-person perspective pull the shooter *into* a field of play for immersive effect. It felt especially so within the visual context of early 1980s video gaming, which had patterned players' eyes via the simple back and forth of *Pong* (1972), the rigid columns of *Space Invaders*, and the right angles of *Pac Man*'s maze.

Battlezone's relatively complex algorithm and immersive display attracted the U.S. Army, which contracted with Atari in 1981 to modify the game as a trainer for its Bradley Infantry Fighting Vehicle. This early example of collaboration between the digital entertainment industry and the U.S. military produced some controversy and resistance. Ed Rotberg, the designer of *Battlezone* and an early star in the industry, refused initially to work on the modification. "I didn't think it was a business that we should be getting into," he later commented. "You've got to remember what things were like in the late 1970s, and

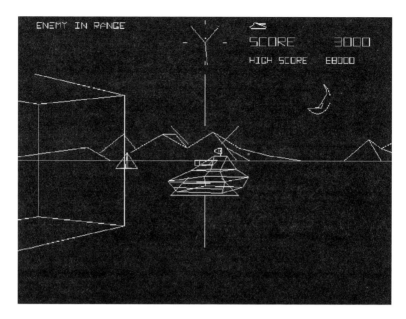

Figure 3.1. The popular Atari arcade game *Battlezone* (1980), using wireframe vector graphics, was hardly "immersive" by early twenty-first-century standards. It nevertheless drew the attention of the U.S. Army, which had Atari design a version for target training for gunners on the Bradley Fighting Vehicle. The object of the game is to maneuver one's tank and its gun turret to shoot enemy tanks and saucer-shaped UFOs. (Atari Games. All rights reserved)

where those of us who were in the business came from—our countercultural background," he explained: "Those of us who found our way to videogames . . . didn't want anything to do with the military. I was doing games; I didn't want to train people to kill."[32] The ethical quandary inherent in wresting killing prowess from play comes through in a 1983 speech President Ronald Reagan gave to a large group of children at Walt Disney World's Epcot Center. His view of the near military future bears a chilling likeness to Orson Scott Card's dystopian state that warrants unknowing commissions of violence among its warfighters:

> Even without knowing it, you're being prepared for a new age. Many of you already understand better than my generation ever will, the possibilities of computers. In some of your homes, the computer is as available as the television set. And I recently learned something quite interesting about video games. Many young people have developed incredible hand, eye, and brain coordination in playing

these games. The Air Force believes these kids will be outstanding pilots should they fly our jets. The computerized radar screen in the cockpit is not unlike the computerized video screen. Watch a 12-year-old take evasive action and score multiple hits while playing Space Invaders, and you will appreciate the skills of tomorrow's pilot. . . . What I am saying is that right now you're being prepared for tomorrow in many ways, and in ways that many of us who are older cannot fully comprehend.[33]

Nor could the youth in Reagan's audience fully comprehend the budding relationship between their video game playing and national defense. Ender understood that he was being trained for battle against the Buggers but mistakenly believed that he retained dominion over determining when and how he would actually fight and kill. As in Card's novel, and to emphasize, the U.S. military would seek to break down the difference between simulation, play, and real war, with the allure of combat video gaming central to the enterprise.

Collaboration between digital entertainment and the U.S. military didn't take on systemic form until the mid-1990s, after the sudden end of the Cold War, when the Pentagon had to follow new cost-cutting directives, and amid a major growth spurt in commercial video-game research and development. A historic marker for the military-entertainment complex came in 1996, when the National Academy of Science (NAS) hosted an extensive workshop in Irvine, California, to explore the opportunities for long-term cooperation between video-game makers, computer science, Hollywood, and the defense industry. Great "differences in culture and business practices" would need to be "overcome, if not altered," the conference leaders acknowledged; the time is now, though, for creating "structures . . . to facilitate collaboration and to allow greater sharing of information between the [entertainment and defense] communities." From the workshop presentations and breakout sessions came a lengthy summary document, *Modeling and Simulation: Linking Entertainment and Defense* (1997).[34]

Widely circulated in the two converging industries, *Modeling and Simulation* became a basic blueprint for the following decade's first-person shooter combat video games. More complex computer-generated enemies—including characters with adaptive behaviors—would provide greater challenge and satisfaction in play while also developing the tactical and operational skills of the warrior in training. Future military simulation would need to achieve the quality of *virtual presence*: "the subjective sense of being physically present in one environment when actually present in another environment." Military researchers believed that virtual presence would be essential for patterning killing. While "creating this sense of presence is not well understood at this time," one central benefit would most likely be "providing an overall experience similar enough to the real world that it elicits the conditioned or desired response while in the real world." As combat video gaming depended on reaching down to the senses and feelings of recreational and military players, "skilled storytelling techniques" would "help participants learn to make the right decisions and take the right actions." Along with such situational details as the geopolitical cause for fighting, "the Department of Defense may be able to learn additional lessons from the entertainment industry regarding the types of sensory cues that can help engender the desired emotional response."[35]

In addition to the seminal document *Modeling and Simulation*, the NAS workshop led to the 1999 founding at the University of Southern California of the Institute for Creative Technologies (ICT), whose explicit mission "is to build a partnership among the entertainment industry, Army, and academia with the goal of creating synthetic experiences so compelling that participants react as if they are real." After an initial five-year investment of $45 million, the Army renewed its support in 2004 with a $100 million grant. ICT has made good on its promise of "engaging, new, immersive technologies for learning, training and operational environments."[36] Its Flatworld project pushes virtual reality to the science-fictional point of *Star Trek: The Next Generation*'s Holodeck by combining digital technology and Hollywood set design to create a life-size "multisensory experience in a space that lets the user move and interact freely and naturally." So impressed with Flatworld technology, the usually tight-fisted Marine Corp paid $1.3 million in 2007 to have it added to its Expeditionary Force Battle Simulation Center at Camp Pendleton, California, and the Rifle Integration Facility near Quantico, Virginia.[37] The Marines would come to call the systems Infantry Immersion Trainers, as of 2013 at Camp Pendleton and Camp Lejeune, North Carolina, which continue to be implemented as Holodeck-like decision houses for riflemen and small-unit leaders.[38]

The Marines had started integrating video games into training at an early date, in the mid-1990s, when its newly formed Modeling and Simulation Management Office (MSMO) in Quantico evaluated approximately thirty commercial games for their promise in helping riflemen improve their "tactical decision-making capabilities." While no one game offered a "robust simulated combat environment," MSMO's Computer War Game Assessment Group produced a *Wargames Catalog* of satisfactory commercial video games. It also recommended that first-person shooters be loaded onto Corps computers for play during on-duty hours.[39] General Charles C. Krulak, commandant of the Marine Corps, implemented that recommendation in April 1997, thereby superseding "previous policy which prohibited playing games on Marine Corps computers." Marines Corps Order 1500.55 states, "All commanders and staff supervisors have a fundamental leadership responsibility" to develop "daily Military Thinking and Decision Making Exercises." Such exercises could take on several forms: the *Marine Corps Gazette* and *Leatherneck* magazine "provide a wide selection of articles and vignettes to stimulate warfighting discussions"; and "commercial books and board-based games provide excellent scenarios for this program." But Krulak's order privileged video games. "The use of technological innovations"—especially computer-driven wargames—"provide great potential for Marines to develop decision-making skills, particularly when live-training time and opportunities are limited." In a "Resources" section, the order refers commanders and staff supervisors to the *Wargames Catalog* for "commercial computer wargames" and notes that MSMO "is developing additional customized and original wargames."[40]

As of 1997 the Marines had "customized" one combat video game for training: *Doom* (1993), the notoriously bloodthirsty first-person shooter that Eric Harris and Dylan Klebold played before the Columbine killings. The National Research Council's report *Modeling and Simulation* speaks highly of *Marine Doom* and, more generally, the Corps' idea of modifying commercial games.[41] Having dropped the original game's intergalactic setting, *Marine Doom* (1997) "trained for cooperation" among four riflemen in a fire team, as video-game critic Ed Halter describes the modification: "The teams could set up fields of fire, conduct flanking maneuvers, and enact 'leap frog' tactics in which some members pin down an enemy with gunfire while others advance ahead."[42] Recognizing the value of such tactical training and decision making in virtual reality—and spurred on by *Modeling and Simulation* as well as the Army's early efforts to produce what would become *America's Army*—the Marine Corps called for bids from private firms for co-developing a first-person shooter video game focused on a four-man fire team engaged in urban combat. The Destineer Corporation eventually won the contract, with *Close Combat: First to Fight* the result of its collaboration with the Marines.

Set in civil-war-torn Beirut, Lebanon, in 2006, with the player's mission to clear the streets of radical Islamist insurgents, *First to Fight* benefits from the considerable expertise the Marines contributed to the game's production, including thousands of pages of Corps doctrine and training manuals. While spending only $700,000 on *First to Fight* (approximately a tenth of the cost), the Marines arranged for more than forty riflemen from the Third Battalion, First Regiment recently back from fighting in Iraq to sit down with designers and describe the particular sights, sounds, and challenges of urban combat. The Destineer Corporation designers tried to put the player in the combat marine's place. As the leader of a four-marine fire team that deploys the 360-degree security doctrine of "Ready-Team-Fire-Assist," the player knows that he or she is engaged in the same patrol tactics "now used in Iraq and Afghanistan," as the game's website emphasized: "When you cross an intersection, ascend a staircase, and engage the enemy, your team will do it the way Marines are doing it, right now, in the most dangerous places in the world."[43]

From Combat Films to Video Games

Close attention to *First to Fight* begs a question: just what sort of reality did the game makers try to copy? It opens into an attractive cityscape of Beirut with a golden mosque dome at its center. A shot pans high across rooftops and streets, then pulls back across several blocks and through a Mediterranean-style balcony and doorframe. The accompanying audio track begins faintly with the beating of a heart and the streets; it shifts, with the pull back, to sonorous Muslim prayer amplified through still, early-evening air. Having established a sense of both place and calm, the game bursts into a staccato of urban combat clips: residents scurry for cover in near darkness; a black-bearded, large-armed fighter launches a rocket-propelled grenade through a market square; and a fiery explosion rocks a café, as orders are barked in guttural Arabic.[44]

Against this backdrop of exotica, tumult, and violence comes the familiar frame of a television newscast and CNN-based music signifying an important update. It's "Battle for Beirut" on INN, International Network News, covered by Amber Wolff embedded with Echo Company of the Marine Air-Ground Task Force. More civilians were killed in the past few hours, she reports, in fighting between Colonel Akhbar al'Soud's Lebanese Militia and "radicals" from the "extremist group Atash," led by "fundamentalist" Tarik Qadan. Photographs of al'Soud and Qadan appear in the playing card form used for marking high-value targets in Operation Iraqi Freedom. "In response to the renewed hostilities," Wolff continues, "U.S. Marines have stepped up patrols to secure the El Karantina district of Beirut." Syrian troops have been identified fighting alongside Lebanese Militia, she adds, thereby increasing the danger faced by U.S. Marines. "This is Amber Wolff reporting."[45]

First to Fight's opening is customary for early twenty-first-century combat video games in its highly stylized composition and in its geopolitical justification for the simulated killing to follow. "Operation Preserve Peace," a joint United States and NATO "intervention" in a Lebanese civil war, has the "Marines Back in Beirut" the lead caption within the INN newscast. Innocent lives are being taken. Geopolitical balance of power is at stake. The mission's drama and significance grow through its cinematic framing. The zoom-out establishing shot of Beirut, the television newscast documenting the reason for committing U.S. troops, the closely edited video-clip sequence that pulls the viewer into the fighting—these are all conventions of Hollywood movies. Indeed, the game's designers borrowed freely from leading late twentieth-century combat films. "*Saving Private Ryan* and *Black Hawk Down* were the most influential," president of Destineer Corporation, Peter Tamte explained that "we went to them for their emotions and experience and scenarios."[46] The first scene in *First to Fight* combining the mosque dome with atmospheric prayer draws heavily from Ridley Scott's overview of Mogadishu, Somalia, in the beginning of *Black Hawk Down* (2001). At the turn of the twenty-first century, even as Americans had started to spend more money on video games than at movie theaters, the older film medium still enjoyed primacy as the most common cultural denominator for seeing war.

While commonplace to recognize Hollywood's influence on video games, it's worth considering the particulars of that carryover, especially in combat games that are used for military training as well as for entertainment.[47] In looking at *First to Fight*, the most basic element taken from the combat film is the choreographed action sequences that, in the newer medium, produce the challenge and action so central to playing the games. "The primary part of what we're doing," as one *First to Fight* designer comments, "is capturing the enjoyable aspect of fighting."[48] To accomplish that goal, many designers looked to specific war films in staging their texts' interactive combat scenes. The well-known indoor firefight scene from the original *Star Wars* (1977) in which Luke Skywalker, Han Solo, and Chewbacca rescue Princess Leia from the Deathstar prison served as a model for close-quarter shootouts in virtual Beirut. Most basic to *First to Fight* are *Black Hawk Down*'s intricately composed battle scenes in which U.S. troops wind their way through the maze-like streets of Mogadishu where almost every turn brings a confusing mix of unarmed civilians and deadly guerrilla fighters.[49]

Combat video-game designers have also drawn from war movie soundtracks and the Hollywood convention of trying to spark certain feelings through audio cues. The audio track of *First to Fight* is "very, very intentional," Peter Tamte has said, referring to its attempt both to recreate the sounds of urban combat that the marines remembered from Iraq (e.g., bullets ricocheting off of cars) and to formulate a player's emotional register during peak moments of patrol and battle.[50] Approximately a quarter of the way through *First to Fight*, for example, the Marine fire team enters a town square and approaches the door of an apartment building thought to hold enemy fighters. Going through doors "in a combat setting can be a hazardous action," the *First to Fight* Instruction Manual explained: they "are a natural chokepoint to use in an ambush."[51] As the player's avatar turns the corner into the doorway, the audio overlays a distinct inside-the-mind sound of dripping water, very similar to the audio byte from the Vietnam War film *Full Metal Jacket* (1987) when Joker is about to open the lavatory door to find the deranged Pyle with his fully loaded M-16 rifle. The sound's effect in *First to Fight* is immediate and uncomplicated. The water-drip sound heightens the suspense of what waits inside the building; its intricacy dramatizes by sharp contrast the following blare of gunfire and shouting.

Despite the significant carryover from Hollywood to video games, at least one quality of film could not yet be extended fully to the newer medium: facial expression. In drawing the *First to Fight* characters, the game designers used photographs of marines from the Iraq War.[52] On screen, these images change and move in only the most rudimentary fashion. *First to Fight* creators sought "emotional captures"—that is, ways to instill in the player the feelings of fear, ferocity, pain, and satisfaction so central to real battle experience and the combat film genre. Audio cueing of emotions went part of the way, but turn-of-the-century technology didn't allow game producers to work with actors, direct them for desired emotional effect, and transfer the subtleties of facial expression into the video text.[53] In terms of evoking emotion, Peter Tamte has acknowledged flatly, "Hollywood has the greatest trick in the book: show the face."[54] Put differently, facial expression marked a major deficit in combat video gaming and its function of calibrating eyes for battle. Excluded is the visage, say, of a marine charging the enemy—his jaw set in grim and deadly determination.

Also missing from video games are close-ups of other fighters observing the action. Combat films have typically modeled both heroism and admiration of such behavior. In *Sands of Iwo Jima*, Sergeant Stryker's squad is pinned down on the beach by heavy machine gun fire from a Japanese pillbox atop a tall dune. A private takes a large dynamite pack and starts up the hill amid a rain of bullets; the camera cuts to a close-up of John Wayne, his face full with admiration and concern; after the marine is shot, the camera returns to Wayne and his now remorseful expression; Wayne then grabs the dynamite and, with his wide-eyed men looking on, makes it to the top where he blows up the enemy position.

A scene from Stanley Kubrick's *Full Metal Jacket* further illustrates the power of facial expression in combat films. Joker has joined a Marine platoon in the battle of Hue. The unit fights slowly through desolate urban rubble, taking casualties while seldom spotting enemy fighters. The view is usually a long shot, from behind the marines, looking past them toward smoldering bombed-out buildings and hidden North Vietnamese Army (NVA) positions. The camera moves closer to a marine, Crazy Earl, who is looking down to reload his M-16. Three NVA soldiers run laterally, from right to left, between two buildings, about one hundred yards away. Although the three are hidden again, Crazy Earl raises his rifle, aiming where they had just run. The camera shot is now over the shoulder—the view follows the barrel of Crazy Earl's gun. Two more black-clad figures start running from one building toward the next when Crazy Earl fires and hits both of them. Cut to a close-up of Crazy Earl. His expression changes from wonderment to beaming unabashed pride over his prowess, and then to a more serious look, one of solemn respect for his accomplishment and perhaps for the men he has shot.

Paradigmatic of war pornography, this ten-second scene from *Full Metal Jacket* anticipates the player's perspective in *First to Fight* and other first-person shooter video games—over the shoulder, that is, with part of the gun's barrel and tip directing the player's line of sight. Video-game designers may very well have been influenced by the stark drama and clarity of Kubrick's point-of-view shot. After *Saving Private Ryan* and *Black Hawk Down*, combat game designers cite *Full Metal Jacket* as most influential.[55] From Crazy Earl's perspective, with gun drawn and set on the spot where NVA targets had just been in fleeting view, one eagerly anticipates the next enemy figures to appear on screen. And after hitting two fast-moving human targets, Crazy Earl then models subjective satisfaction from fighting and killing in combat.

Having considered the congruencies between combat video games and Hollywood war films—and having established the combat film genre's particular capacities for exciting viewers—the question becomes what did the newer medium add to the turn-of-the-century prospect of motivating Americans to fight and kill in battle? Rapid advancement in digital technology made video games more affordable and adaptable than feature-length films. The $5 million the U.S. Army paid for *Full Spectrum Warrior* in 2005—not to mention the $700,000 the Marines spent on *First to Fight*—paled in comparison to the $100 million price tag of an in-house recruitment film; the ability to easily feed in geospecific data to a game design during a time when the U.S. military started to recalibrate itself for urban warfare made video gaming all the more attractive.[56] Less certain has been the neuroscientific basis for the military's increasing investment in combat video games and simulation training. Defense-industry psychologists as well as digital media theorists have concentrated, though, on video gaming's interactivity: physical manipulation of the controls, and the corresponding bodily action represented on screen, produces muscle memory or "procedural memory" that can be recalled in actual battle.

The Value Added to Military Training

The first decade of the twenty-first century brought a sharp rise in the use of video-game simulation for instruction, social interaction, and therapy. Surgical training and quit-smoking simulation programs joined military and civil aviation trainers. Video games helped firefighters prepare for terrorist attacks. The cable television network MTV extended its *Real World* (1992) program into *Virtual Hills and Laguna Beach* (2007), where "You can not only watch TV, but now you can actually live it." In January 2010 the "simulated reality" site *Second Life* (2003) had 18 million registered accounts. Racecar driver Denny Hamlin won the Pocono 500 in 2006 by preparing almost exclusively on a video game. With every turn and tree on the real Pocono track included in the program, Hamlin explained, "they've got everything mapped out perfect[ly]. Visually, I know where my [acceleration] points are," and "it helps with track awareness."[57] Within two years of the U.S. invasion of Baghdad, USC's Institute for Creative Technologies had designed *Virtual Iraq*, a prolonged exposure therapy system for soldiers and marines suffering from post-traumatic stress syndrome.[58] The post–September 11 wars have seen American troops playing video games before, during, and after deployment.

A conceptual basis for this widespread development of goal-directed video games and simulation programs is a thoroughgoing cognitivism. The brain can be seen as a vast information processing system, whose memory can be uploaded with intent and purpose, since that stored data, in responding to environmental stimuli, informs individual and social behavior, including wartime fighting and killing. For some psychologists working on military training, the quality of human consciousness becomes all but lost in the equation. Consider a paper delivered by four defense-industry psychologists in 2006 in at the Interservice/Industry Training, Simulation, and Education Conference (I/ITSEC), Orlando, Florida (another hub of the military-entertainment complex). Entitled "Harnessing Experiential Learning Theory to Achieve Warfighting Excellence," the paper focuses on enriching simulation ("complex virtual environments") for optimal effect in battle during the Afghanistan and Iraq wars:

> As we prepare warfighters for challenging, uncertain situations, we recognize that the richer their databases, the more available mental resources they will have to apply or reconfigure in response to dynamic, unpredictable situations. We contend that experiential learning should be designed to fill those databases with accurate representations and awareness of relationships among them to promote ready access. Experiences should be structured to develop principles and procedures of the discipline, and to optimize the interactions between working memory (the workspace that processes new information and attaches it to previously learned information) and long-term memory, where information and representations are held permanently.[59]

Not all cognitivists spoke in such full-blown computational terms. "It's fascinating now what we're learning about the human brain," Michael Macedonia commented straightforwardly in an interview at the 2004 I/ITSEC conference: "We are our memories. . . . So a lot of what we're trying to do in training is creating memories. Memories that last forever." The computer metaphor did creep in, though, as Macedonia continued: "In the training world, we try to create a virtual reality in the virtual reality program in our head. We are trying to mess with that program a little bit," he concluded, "so you remember long afterward that experience."[60]

What emerges in these comments from the I/ITSEC conferences and from early twenty-first-century military simulation overall is a strategy of creating virtual veterans, healthy and sound new marines and soldiers who also carry a store of memories from simulated combat experiences designed to make them less hesitant and more deadly fighters.[61] The memories themselves— their type as well as content—are crucial to understanding what video gaming may have done to ready American troops for battle. In specifying the type of memory video gaming induces, one significant difference is that first-person shooter combat games don't key on watching another figure but rather enact behaviors in response to images. The decision-making quality of video gaming is emphasized in *Modeling and Simulation*, the summary report from the 1996 National Research Council workshop joining defense to entertainment. "Simulation, VR, videogames, and film share the common objective of creating a believable artificial world"—what the document describes as "the perception that a world exists into which participants can port themselves and undertake some actions. In film this process is vicarious," *Modeling and Simulation* states crucially: "in simulation, VR, and gaming it tends to be active." It "allow[s] people to directly perform tasks and experiments much as they would in the real world . . . people often learn more by doing and understand more by experiencing than by simple nonparticipatory viewing or hearing information."[62]

Modeling and Simulation puts it a bit too simply. The notion that film viewing is "nonparticipatory" wouldn't sit well with the generation of film scholars who have concentrated on audiences' interpretive agency and widely subjective experiences. And the range of activity in video gaming is easily overstated, especially within the combat genre at the turn of the century. Grounded in an "if this, then that" algorithm, a first-person shooter's action is delimited by the game's pathways, the player's experience predetermined by narrative fiat and search-and-destroy directive.[63] It would have been more accurate to label video gaming's activity as interactivity. Video gaming's particular feature of registering and responding to decisions made through neuromuscular action—manipulating a joystick to move on screen, for instance, pushing a button to shoot— does entail interactivity of a more material kind than viewers' identity formation and other psychological engagement with film.[64]

Video gaming is a form of play. Immersion, that "being-there" sensibility so sought after in military simulation, is felt in confrontation with the enemy— that is, while engaging the computer's artificial intelligence as if *somebody* (rather than something) is on the other side. Video gaming's reality effect is as much emotional as geographic. It's about competition and performance under pressure. "It gives you a sense of reality," USMC Corporal Justin J. Taylor exclaimed in 2003 while training on a first-person shooter combat video game. "You get that nervous feeling: do you really want to go around the corner or not? You want to complete the job you've been assigned to."[65] Corporal Taylor undoubtedly went around that corner on screen, an action that, by the logic of simulation, would have made him more likely to round corners once deployed overseas: memory of the first combat experience—virtual, without risk of bodily harm—informs and augments behavior while fighting in real place and time. "We have a term in virtual reality we call presence," as Michael Macedonia has explained with characteristic plainness: "that's what we're trying to do with these soldiers with these training systems. When they go out there, they'll think, I was here. I've been here before. I know what I'm supposed to do. And that's the essence of it."[66]

It should be noted that the physical interactivity of early twenty-first-century military simulation has been limited by video gaming's keyboards and modular control systems. More promising for instilling procedural memories of combat is technology like Nintendo's Wii and Xbox's Kinect that strives for one-to-one correspondence between the bodily action of player and avatar. Virtual reality headsets provide fuller range of player motion and perception, including walking through three-dimensional spaces and 360-degree vision. While the Marines' Infantry Immersion Trainers are an interim step toward full-body virtual reality training, Holodecks may very well complete the continuum of pulling simulation and actual fighting so close together that one has little reason to differentiate between the two.

If joysticks, button pushing, and HD screens pale in comparison to the bodily interactivity of full holographic simulation, it should also be recognized that they made a significant difference from the pop-up targets developed for post–World War II rifle ranges. Turn-of-the-century combat video games, designed for both training and entertainment, took on a practical purpose in readying American troops to fight in the post–September 11 wars—conflicts far-removed from U.S. soil and fiercely opposed both abroad and at home. "It felt like I was in a big videogame. It didn't even faze me, shooting back," is how U.S. Army sergeant Sinque Swales described his first firefight in Iraq and the way playing *Halo 2* had prepared him for that combat. While on patrol in Mosul in 2005, his unit suddenly encountered a group of Iraqi insurgents: "It was just natural instinct. *Boom! Boom! Boom! Boom!*" Michael Macedonia would undoubtedly have been happy with Swales's response in Mosul and with his account of video gaming's déjà vu effect. Combat video gaming makes young soldiers and marines "feel less inhibited," as Lieutenant Colonel Scott Sutton, head of USMC training and technology, put it in 2006. Its simulated battle experience is stored and recalled "down in their primal level," and that "provides a better foundation for us to work with."[67]

Another idea to consider about the type of memory instilled through video games involves bodily register. The "knowing what to do" feeling from simulated combat may cohere through what psychologists and memory studies scholars call procedural memory. Compared to "declarative memory," which involves storing and recalling "information about events, people and objects . . . through symbols and representation," as historian Karen Till writes, procedural memory is the process by which "experiences are encoded directly into neural programmes." Procedural memory bypasses linguistic and other symbolic systems of representation, record, and recall altogether. It's retained through bodily cognition.[68] Under this view, military simulation via first-person shooter combat video gaming doesn't so much craft a mind's eye for battle as it sets muscle memory for fighting. It is less similar to Hollywood war pornography than it is to athletic competition and sports' long-recognized effectiveness in preparing men for battle.

Video gaming's effectiveness in preparing its players for war can be broken into two parts. First, simulation: video games are like "Combat Town" at Twentynine Palms, California, and other Middle Eastern city mock-ups the Marines and Army constructed stateside for predeployment urban combat training during the Iraq War; often featuring live-fire exercises—and sometimes including native role players meant to complicate fighting amid civilian populations— these mock-ups instilled memory of battle among individual troops and their units before they got to Iraq; both video gaming and the combat towns worked via cognitive embodiment and muscle memory by creating active rehearsals of the sights, sounds, behaviors, and decision making involved in urban warfighting. Second, video games are played. Designed for entertainment value as well as for simulation, video games transcend teaching players the cognitive skills for negotiating urban combat via their competitive and agonistic essence that is still so central to war itself. Video gaming may be to combat today as football was to combat one hundred years ago.[69] Both video gaming and sport instill a spirit in their players conducive to fighting and killing in combat.

Fighting in the Digitized Streets of Beirut

The Marines and the U.S. military tried to realize many practical benefits from turn-of-the-century combat video games. While more and more recruits started boot camp as virtual veterans, the armed forces followed the National Research Council's 1996 finding that "advances in information technology" had made "modeling and simulation a cost-effective alternative to live training."

The *Modeling and Simulation* report added momentously that "changes in the geopolitical environment are requiring the military to plan for actions not only in traditional regions of conflict" and that the Department of Defense "needs to be able to rapidly model varied locations and scenarios to assist in training troops."⁷⁰ Combat video games would prove to be most adaptable. The Army and Marines continued to use live training in its many forms, from rifle-range target practice to whole squads and companies preparing for counterinsurgency assignments by patrolling imitation Afghani villages in stateside Combat Training Centers. Such centers did not come cheaply or easily, however. Each training session at Fort Polk in Louisiana, for instance, involved more than two thousand logistics experts, writers, role players, and pyrotechnists for setting explosions.⁷¹ Digital simulation, by comparison, has been far less labor intensive and could be widely disseminated to troops for integration with other types of training.

Combat video gaming's adaptability—its emergent capacity for "rapidly model[ing] varied locations and screenings"—was well timed with the late 1990s shift by the American military toward preparation for urban warfare. With the Cold War's end having already initiated fundamental reappraisal of U.S. armed forces, transformation became more imperative after the 1993 battle of Mogadishu, where U.S. Marine and Army Special Forces had been sent to alleviate the famine caused by Somalian civil war. As chronicled in Mark Bowden's book *Black Hawk Down* (1999), several hundred Somali militia and eighteen American troops lost their lives in overnight small-arms conflict fought through the city's narrow streets. Set amid a large civilian population, the combat demonstrated the limitations of American military effectiveness, even when its forces enjoyed far superior communications, technology, and firepower.⁷² Further demonstration of so-called asymmetric urban warfare came the following year, when Chechen rebels destroyed more than 100 of the 120 Russian tanks inside the city of Grozny. "The open battlefield we prepared for in response to the Soviet threat is the least likely type of warfare as we enter the 21st century," USMC lieutenant general John Rhodes stated in 1999. "Instead," he continued while testifying before the Senate Armed Services Committee, Emerging Threats and Capabilities Sub-Committee, the "sprawling urban areas of the world, with their sea of noncombatants and maze of streets and buildings, appear to be the most likely future battlefields."⁷³ Overarching concern about urban warfare provided the most immediate context for Destineer Corporation's creation of *Close Combat: First to Fight*. Preparation and training after September 11 for fighting in Kabul and Baghdad and other Afghani and Iraqi cities only added to its urgency.

Destineer and the Marines agreed to make *First to Fight* in 2001, just after the Corps' Operation Urban Warrior, developed by the USMC Warfighting Laboratory (established in 1995). Part of a Five Year Experimentation Plan, Operation Urban Warrior worked from external estimates that by 2020 roughly 70 percent of the people in the world will live in cities, and, as General Rhodes said before the Senate, "70% of these cities will be within 300 miles of the coasts—the traditional operational area of Marines and naval forces."[74] Operation Urban Warrior incorporated the Somalia and Chechnya experiences, emphasizing that combat in cities "will involve maneuver and close range engagement in an environment characterized by concentrated cover, concealment and obstacles." Urban warfare is "messy, entangled and chaotic," a 1998 operation concept paper emphasized: it "offsets many of the strengths in the traditional American way of war." Culminating in a large-scale practice invasion of the San Francisco Bay Area in 1999, Operation Urban Warrior produced several adjustments for fighting in closed and disconnected city spaces. First among these recommendations was a "downward shift" in the "locus of decision making." The four-marine "squad is the best maneuver element in the urban jungle." And, with that finding, the squad leader will "become the lowest level battle leader capable of independent operations on the constrained urban battlefield."[75]

Operation Urban Warrior's recommendation to concentrate on low-level decision making became a centerpiece of new Marine doctrine through General Charles C. Krulak's widely disseminated article "The Strategic Corporal: Leadership in the Three Block War" (1999). Published during Krulak's last year as Corps commandant, the essay picks up from his 1997 order authorizing on-duty video play for improving "military thinking and decision making" among all marines. "The Strategic Corporal" draws directly from Mogadishu; *Black Hawk Down* is "essential reading for all marines." Beginning with a fictitious Marine deployment named "Operation Absolute Agility" set in "the war-torn capital of Orange—a Central African nation wracked by civil unrest and famine," the essay centers on Corporal Hernandez, squad leader in Second Platoon, Lima Company, who is suddenly faced with a growing mob chanting anti-American slogans; a downed U.S. helicopter; three vehicles filled with heavily-armed rebel militia; a wounded marine; and an INN news crew recording all of the unfolding action. "Corporal Hernandez was face to face with the grave challenges of the *three block war*," Krulak wrote, referring to urban operations' overlapping missions of conventional combat, peacekeeping, and humanitarian aid: and the corporal's "actions, in the next few minutes, would determine the outcome of the mission and have potentially strategic implications."[76]

Krulak's answer to the predicament of such responsibility resting on the shoulders of young low-ranking marines was to rebuild the Marines around a force of "strategic corporals": recruit "bold, capable, and intelligent men and women of character"; use boot camp to instill "the Corps' enduring ethos" of "honor, courage, and commitment"; and realign combat training and other instructional programs to "aggressively cultivate" individual traits of "decisiveness, mental agility, and personal accountability." It was this charge—distilled through the Corps at the very turn of the century—that most closely informed the production of *First to Fight*.[77] "In the end," Peter Tamte explained in the online magazine *Wargamer*, "our goal in making *First to Fight* is to get you as close as we can on a computer or an Xbox to being a Marine fire team leader in modern urban combat" (fig. 3.2). Carryovers from "The Strategic Corporal" to Tamte's game are fundamental and explicit. "The values of honor, courage, and commitment" are highlighted several times in the Instruction Manual, as are tributes to Marine character and tradition: "They are our nation's noble and elite warriors," for example.[78] The game's first image is a dark screen with an axiom slowly appearing in light-gold lettering: "The greatest supporters of peace are those who are sworn to risk their lives when war occurs. General Charles Krulak, 31st Commandant, USMC." *First to Fight* even borrowed the fictitious INN cable news network from the general's country Orange scenario.

The interface between the Marines Corps and Destineer was multifaceted. The several "groups of active-duty marines, just back from combat in Iraq or Afghanistan" received the most publicity, as Tamte summarized proudly in *Wargamer*. They are sent "to our offices to sit down side-by-side with our engineers, artists and designers," he continued, "spend[ing] many days during each trip with our team to put the exact tactics they used during combat into *First to Fight*." From the perspective of military simulation, the concept is impressive: a tight loop between recent combat experience, selective memory of that experience, and transference of that memory into the video game's content, which in turn creates virtual combat experiences for players, some of whom will be deployed to the site of the original combat to reenact the rarefied set of

Figure 3.2. A screenshot from first-person shooter *Close Combat: First to Fight* (2005). Producer Destineer Corporation and the U.S. Marines agreed to develop a combat video game that concentrated on MOUT (military operations on urban terrain). General Charles C. Krulak's influential article "The Strategic Corporal: Leadership in the Three Block War" (1999) provided the military doctrine upon which the game was built. The player assumes the role of a Marine corporal, leader of a four-marine fire team clearing the streets and buildings of insurgents in Beirut, Lebanon. (Destineer Corporation. All rights reserved)

behaviors. Procedural memory is the goal. "They personally demonstrate these tactics to us as four-man teams," Tamte elaborated; "they painstakingly detail their formations and fire-sector coverage on whiteboards."[79] Next, Destineer used a motion capture system for transferring the tactics into the game: with their limbs and joints marked with reflective sensors, hired actors ran through the Ready-Team-Fire-Assist maneuvers against a green screen, their physical actions tracked and recorded digitally. While emotion capture of facial expression is still rudimentary, Destineer could rely on recording, as a *First to Fight* designer put it, "the kinetics of combat" with precision, and then have players absorb that motion in ideal form through the game's experience.[80]

Along with battle-hardened marines visiting the Destineer studios, exchange between the two happened at a more basic level: a certain institutional memory passed from the Corps to the game maker. The Marines assigned "Subject Matter Experts" to the project; Destineer production teams visited

Marine bases and studied thousands of pages of warfighting doctrine. The Marine Corps issued a "comprehensive Statement of Work describing in great detail all the functionality we were required to provide in the product," as Tamte explained; those responsible for generating this contractual document were "five to six levels down from General Krulak."[81] It's at this level where Operation Urban Warrior criteria passed into *First to Fight*. An urban setting had been agreed upon from the start.[82] After the September 11 attacks, a Middle Eastern city became imperative, though for this first Marine-based game Destineer and the Marines shied away from locating it in Iraq or Afghanistan. Beirut's recent history of religious conflict and war made the large Lebanese coastal city quite plausible for *First to Fight*'s scenario of Marine intervention. The choice also returned the Marines to where in 1983 two suicide bombers had killed 220 of its men.

Destineer designed *First to Fight* game play around the tactical decision making and violence required to kill armed forces positioned among a civilian population and within a foreign city's diverse topography and infrastructure. Operation Preserve Peace is really a clearing action. "If you kill too many civilians," as the Instruction Manual reads, "your mission will be over." And more points are scored if one captures rather than kills the five faction leaders (i.e., the "High Value Targets").[83] Otherwise, there's no incentive to hesitate before shooting. The opening INN cable news report—a quick summary of the strategic situation before shifting to Amber Wolff embedded with Echo Company—ends with a deep baritone voice over: "And, now, the U.S. Marines prepare for an aggressive assault into the heart of Beirut." *First to Fight*'s narrative tree is quite simple: move through different city sections and into its markets, cafés, hotels, warehouses, and shops and eliminate the opposing forces placed between you and predefined checkpoints. Slums are cleared of militants; a hospital and American University are reclaimed for proper use; militant leaders are pursued and captured or killed; Beirut is returned to its peaceful inhabitants.

A clear carryover from the turn-of-the-century Operation Urban Warrior recommendations has *First to Fight*'s combat unfold on four distinct urban planes. The 1998 Urban Warrior concept paper recognizes that in the "constrained battlespace" of the city, "units will maneuver on four planes: (1) the subterranean plane using the sewers and subways; (2) on the surface plane using the floors of the urban canyons; (3) within the structural plane from building to building, and (4) in the air above the city."[84] In *First to Fight*, one encounters opposing forces in bunkers, atop fortified buildings, and in other entrenched positions that cannot be overrun by the four-marine Fire Team; as leader, though, the player enjoys sniper and mortar support as well as an air strike option—the Marines control the air above; "radio in the coordinates of the targets," the manual reads, "and shortly thereafter, Cobra [helicopter] gunships will arrive, eliminating the OPFOR [opposing force] threat."

Following new warfighting doctrine and instruction from veteran marines, Destineer devoted considerable detail to Ready-Team-Fire-Assist maneuvers through the myriad of buildings in Beirut. Building entryways, stairs ("fatal funnels" to Marines), and doors to rooms all require special tactics, including a team "stacking" formation for "room take downs." (A "Frag and Take-Down" order has the marines grenade the room before rushing inside.)[85] Beirut's "surface plane," where most *First to Fight* action occurs, is indeed canyon-like: except for old cars strewn over the curb, the streets almost always appear empty; and, with large decrepit buildings looming above, the sky, even at midday, is half lit. Finally, in following Marine specifications, Destineer wrote in several forays through the "subterranean plane." Because there is no subway in Beirut, the fire team drudges through the sewer systems' dank passageways, where sludge runs over the marines' boots and OPFOR wait behind undulating concrete walls.

After being built on specifications for urban warfare training, *First to Fight* has been used extensively by the Marines and played by recreational gamers drawn to its "authenticity." *First to Fight* never claimed to offer the fantastic intergalactic combat of *Halo 3* or the spectacular action sequences (such as high-speed snowmobile chases and firefights) of *Close Combat: Modern Warfare*. The commercial game's foundational appeal is that one wants to know or feel what it's like to lead a USMC fire team in modern day (read Iraq and Afghanistan War) urban combat. "We believe authenticity IS fun," as Tamte wrote in his "Developer's Diary." He continued: "An incredible thing about videogames is that they allow us to get a taste of someone else's really dangerous life from the safety of our own living room."[86] Online responses to the *First to Fight* documented its closeness to actual combat tactics. In chastising a player for complaining that there's "no space to operate inside buildings," a former marine wrote in 2005, "no kidding genius, it's called as close to real life as it gets . . . ever been in a hallway with four guys in 782 [standard issue combat] gear? It's kinda hard to operate." Another former marine said online that he "loves" the game, adding that it's "good" on basic tactical doctrine.[87]

Despite its pursuit of authenticity, Destineer decided to "water down" the violence in *First to Fight* so that it could earn a more inclusive TEEN rating. Designers calibrated the game's violence to the level of primetime television action adventure shows such as *The A-Team* (1983–86).[88] In contrast to its moderate graphic violence, *First to Fight*'s structural violence is considerable. This measurement includes its effectiveness in building tactical skills for urban warfare, which the Marines immediately tested and proved by running two groups through a training program; as Tamte reported, "those who had played the game scored far better than those who hadn't" in such categories as "live fire tactical evaluation," "movement techniques," and "tactical awareness."[89] More critically, violence is built into the very ontology of *First to Fight*. Within the game, one exists to clear Beirut's streets by searching for and destroying Islamic extremists. To emphasize, there's little reason to behave differently. A gamer commented online in 2006 that he did have the choice "to shoot or capture" the five factional leaders, "but nobody seems to care one way or the other and shooting them was less of a hassle." Another gamer—one who "personally dig[s] the hell out of this game"—wrote that "I love it when I point my team in a direction and tell them to go kill. I hear'm in there yelling, shooting, and even the occasional buttstroke."[90] Contributing to the game's structural violence is an AI "authenticity" feature that profiles the enemy's morale and discipline. With its algorithm running on an inverse relationship between the enemy's "will to fight" and "great adversity," the game has a built-in bias toward "ruthless destruction," as the Instruction Manual describes, encouraging one to use "massive fire" in carrying out a mission.[91]

Recent history suggests, however, that human will works in a way opposite to *First to Fight*'s premise. For much of the Iraq War, insurgents' will to fight increased with the application of American force. Likewise, the War on Terror may have had the opposite effect of its stated intention. Substantiation for at least the first part of the point can be found in *The U.S. Army/Marine Corps Counterinsurgency Field Manual* (2006), the "paradigm shattering" blueprint for transforming the doctrine and tactics guiding the Iraq War. Coauthored by David H. Petraeus, who would become commanding general of the Multi-National Force—Iraq in 2007, the *Field Manual* lists several "unsuccessful practices" needing overhaul, including "overemphasiz[ing] killing and capturing the enemy rather than securing and engaging the populace." It also proffers new maxims for counterinsurgency. "Sometimes the more force is used, the less effective it is" all but countermands *First to Fight*, a combat video game whose non-Marine characters are either enemies to be killed or civilians to be left alone.[92] By the

94   *Light It Up*

Field Manual's lights, the game shaped dysfunctional marines. With no computation for empathy let alone nonlethal engagement, *First to Fight* extended an inappropriate and deadly project into the minds and bodies—into the play and eye—of marines and civilian gamers.

As for *Six Days in Fallujah*, there's almost an unfairness to it having been shutdown. Was *Six Days in Fallujah* any more gratuitous than, say, the History Channel's *Shootout: D-Day Fallujah*, a television program that, as discussed in the previous chapter, traded on the aesthetics—the pacing, graphics, and perspectives—of combat video games? Peter Tamte, among others, including Marine Corps adviser Mike Ergo, emphasized that it was always supposed to be "a vehicle for storytelling."[93] The decision to make the game a third-person shooter underlines its intended documentary effect. "It's an opportunity," Tamte claimed, "to experience the true stories of the people who fought in one of the world's largest urban battles of the past half-century."[94] In considering its historical or "narratologist" bearing, though, one would also have to ask just how even-handed *Six Days in Fallujah*'s accounting of the battle and its violence would have been? Tamte and other developers touted *Six Days in Fallujah*'s powerful new game engine, a hallmark of which was the destruction one could achieve by blowing holes in houses and knocking down building walls to get at the insurgents. Certainly, in this regard, *Six Days in Fallujah* accurately represented Marine tactics during the November 2004 battle. None of the *Six Days in Fallujah*'s promotional literature or gameplay videos, however, references the hundreds of Iraqi civilians killed amid such destruction. If historical in orientation, the game was celebratory of Marine firepower and heroism in purpose. Like so much U.S.-sourced film, video, television programming, and literature on the Iraq War, *Six Days in Fallujah* omitted some of the most material results of the battle, consequences—unintended or otherwise—that historians will be taking up for many years to come.

What separated *Six Days in Fallujah* from other narratives was indeed its ludology, for surely combat video games incorporate *both* play and narratology. They combine storytelling with the fun and challenge of competition. They entail a type of embodied decision making that isn't experienced in other cultural forms. These distinctive traits offer, as the Marine Corps has realized, tremendous promise for military training and force projection itself. Perhaps the most profound effect combat video gaming has on real warfighting goes to the quality of *presence* that Michael Macedonia and *Modeling and Simulation* found so important in game design. Isn't it possible that the effect of presence in the virtual reality of a video game may lighten the weight of a moment and one's action in actual combat? A First Reconnaissance Marine in the March 2003 invasion of Iraq described coming under attack: "I was just thinking one thing when we drove into that ambush: 'Grand Theft Auto: Vice City.' I felt like I was living it when I saw the flames coming out of windows, the blown-up car on the street, guys crawling around shooting at us. It was fucking cool."[95] Perceptions of "being there" are relative and dynamic entities. As we'll discuss in chapter 5, the issue of ontological detachment is thoroughly exacerbated in flying and killing from Predator drones. But even for the Marine rifleman, what kind of "here" does he inhabit, and what kind of being in that here does he possess, when a basic frame of his perception is ever more immersive simulation and video gaming?

CHAPTER FOUR

Counterinsurgency and "Turning Off the Killing Switch"

In a "Letter of Appreciation" dated June 23, 2005, Lieutenant Colonel Patrick J. Malay, commanding officer of Third Battalion, Fifth Regiment, praised his marines for a job well done under challenging and dramatically changing circumstances. The 3/5 had been a lead infantry battalion in the November 2004 attack on Fallujah. "Then as the Consummate Professionals that you are," Malay wrote, "you turned off the killing switch and turned on to re-population of the city, providing security, humanitarian assistance and ultimately conducting the 30 January Iraqi National Elections."[1] Malay's marines had fought the three-block war that General Charles Krulak described in 1999. With the early 2005 Iraq elections having been the strategic goal of eliminating in November the insurgent hold on Fallujah, 3/5 had afterward taken up patrol operations and rebuilding what it had just helped destroy, while still fighting and killing insurgents trying to reenter the city. "Turning off the killing switch" is hardly a complete description of 3/5's months following the battle for Fallujah, but the phrase does mark a second, conceptually distinct part of the Marines' force projection in the Iraq War: it points to the counterinsurgency that 3/5 and the rest of I Marine Expeditionary Force started fighting after the 2003 invasion, and the notion of Marines doing anything but killing alludes to the "cultural turn" taken in U.S. military doctrine during the early twenty-first century.

Counterinsurgency doctrine's revival in the first decade of the twenty-first century involves a dimension of the military that often goes overlooked amid recruitment, training, and weapons procurement; bureaucratic processing of personnel, logistics and supply needs; and the vast array of other work, structures, and systems incorporated into building armed forces for making war on other people. The U.S. military deals in ideas. It produces knowledge. With its own academies, universities, war colleges, warfighting laboratories, and lessons-learned departments; its own academic journals, presses, websites, and magazines; and its own network of conferences, think tanks, lecture circuits, and speech making—the U.S. military presents the intellectual historian with broad avenues of voluminous research along with the promise of an uncommon causal relationship between ideas expressed, policy implemented, and behavior modified. That's certainly the case with Iraq War counterinsurgency doctrine. Beginning with Krulak's 1999 *Marines Magazine* essay "The Strategic Corporal: Leadership in the Three Block War" and culminating in the U.S. Army and Marine Corps 2007 *Counterinsurgency Field Manual*, attention to caring for the civilian populace as much as killing the enemy would become formal military strategy by the war's fifth year. Before that, though, counterinsurgency doctrine had been heartily embraced by the Marines in Al Anbar Province. "Turning off the killing switch"—or certain variations on that concept—can be traced back in the Marine's history to the "Banana Wars" of the early twentieth century and the Corps' long-standing self-understood role as a colonial force. As that earlier history shows, marines' attempts to reduce their level of force while considering the needs of the people whose land they occupy is a very difficult proposition for the people subject to the occupation and, very differently, for the marines themselves.

The marines' subjective experiences in occupying Al Anbar are part of examining the Iraq counterinsurgency. Interviews with infantry from the very first stage of the occupation show familiarity with the dichotomy between killing and humanitarian assistance. Corporal Nick Poletis, a fire team leader from Kilo Company, Third Battalion, Second Regiment, spoke from Camp Lejeune in December 2003 about his unit's first deployment in the 2003 Operation Iraqi Freedom. After fighting in the March battle of Nasiriyah, 3/2 turned to patrol, aid, and rebuilding assignments. "As Marines we're trained to kill," Poletis commented, "but we also got to flip that switch and do the humanitarian thing." He repeated the point, speaking as if the switch idea had been set in his self-consciousness as a combat marine: "So we got the best of both worlds, we got to see some combat and then we got to flip the switch and do the humanitarian thing."[2] Such matter-of-factness stands in stark contrast to the devastated edge with which many marines speak about the Al Anbar occupation from 2004 to 2007. Discerning insurgent threats amid the citizenry— turning the switch on again, then off, on, off . . . —would be no straightforward matter in a world of foot patrols through Sunni streets, clogged checkpoints, suicide bombings, and ever more sophisticated IEDs (improvised explosive devices). Krulak never said the three-block war would be easy for marines. But the occupation's duration along with the steadily increasing insurgency made Al Anbar one of the most difficult Marine missions since the Vietnam War.

That difficulty can certainly be seen in the combat marines who survived the occupation and who were closely involved in its violence. Corporal Tom Lowry, Third Battalion, Fifth Regiment, served under Lieutenant Colonel Malay in Al Anbar during the unit's second Iraq deployment. Wounded in a Baghdad suicide bombing in April 2003 during the initial invasion, Lowry returned to Al Anbar with 3/5 in spring 2004. "We left it to the Army," Lowry later recalled in a 2012 interview, "and we came back" to "a mess can." Lowry's memory of that deployment—wracked with anguish, guilt, and digression—concentrated on the drastic change in mission. "Marines are there to kill," Lowry stated flatly. Instead "we became police officers," he continued, "which was horseshit. That's not what we're there for. We're there to take out the bad guys. That's it. And get out of there." Lowry's recollection after this point became fragmented and hard to follow. He spoke of female suicide bombers "who pretended to be pregnant." He spoke of checkpoints where insurgents hid behind women and children. He spoke of following orders to "take them all out." What is clear from Lowry is the death he caused and the severe toll the occupation took on him. You can see it on and in his person. And it's plain in his words. "You think I want to do this," Lowry pleaded: "I have nightmares everyday. I can't sleep." And, then, as if directly addressing his former superior officer Malay: "I'm not a light switch. You can't turn me off and on."[3]

This chapter examines the culture clash over the U.S. Marines counterinsurgency in Al Anbar and the Corps' effort to retool itself for humanitarian operations. At one institutional level, the Marines succeeded in infusing, as Lieutenant General James Mattis described, a "keen understanding of culture" into its training and doctrine for "when the enemy hides among innocent people": predeployment exercises at Twentynine Palms and other bases adopted language and cultural awareness drills; the Corps created the Center for Advanced Operational Culture Learning, replete with a civilian staff of newly minted Ph.D. anthropologists; the Commandant's Reading List added *The Arab Mind* (1973) by Raphael Patai along with several titles from the culture-centric canon of classic counterinsurgency literature; and the Marine Corps University Press published *Among the People: U.S. Marines in Iraq* (2008), among several other glossy photograph-filled volumes depicting peaceful human relations in Al Anbar Province. Heading this enterprise was Mattis himself. About the most quotable Marine general since Smedley Butler, Mattis coined the central Marine tenet for the Iraq counterinsurgency: "First Do No Harm—No Better Friend, No Worse Enemy." And as commanding officer of the First Marine Division, Major General Mattis introduced "culture sensitivity" into preparation for the 2004 Al Anbar deployment, publicly criticizing the Army's "iron hammer" response to the Sunni insurgency. By early 2004 the Marine's cultural turn had been absorbed in the combat regiments headed for Al Anbar, with Mattis often the first reference point for that orientation. As Lieutenant Donovan Campbell of the Second Battalion, Fourth Regiment described the clash of cultures within the Marines: Mattis had taken "every step necessary to instill this strange population-centric mind-set into a force oriented toward high-intensity combat against a well-defined enemy."[4]

At another institutional level, the idea of shock troops seeking to "do no harm" is nonsensical. For Corporal Lowry and other marines in the middle of the occupation, "this strange population-centric mind-set" would be unmanageable, an exercise in applied anthropology gone bad or, rather, misbegotten. As counterinsurgency doctrine took hold and expanded in the Iraq War's latter years, senior officers in the Marines would express grave doubts about its excesses and applicability in most operations. Outside the military, among academic social scientists whose work on culture fueled the renewed counterinsurgency doctrine, the notion of the Marine Corps' humanitarianism directed toward an occupied people would be dismissed as, at best, extremely dubious—impossible to achieve, woefully disingenuous, and hegemonic, a projection of force by slightly different means.[5] Within this framework, close examination of Mattis's public pronouncements on friendship with Iraqis renders them incomprehensible. To shift perspective back to the Marines, though, one can argue that Mattis and the Corps understood its predicament and couldn't be expected to do anything else but develop realpolitik measures for fighting insurgency. The real mistake—the historical problem—could be found in the political decision to invade and make war on Iraq in the first place. It was hardly the first time the Marines had been ordered to enforce America's imperial ventures. The Marines had produced *The Small Wars Manual* (1940) from those earlier efforts. A counterinsurgency manual of its own renown, it bears close similarities with U.S. Army general David Petraeus's early twenty-first-century text, while also giving definition to the Corps' intraservice role of handling the nation's "dirty work."

Empathy, General Mattis, and the Profound Paradox of Marine Humanitarianism

Does turning off the killing switch involve empathy? The question goes to the heart of the reservations over counterinsurgency doctrine. In defining empathy as having the feeling of another aroused in ourselves, it's possible to understand that perceptual dynamic and the range of moral sensibilities that often go with it as war's opposite. Set against, say, bloodlust in combat, or instantaneous obedience to an order to fire and kill in battle, empathy highlights the self-reflexivity that imagines the consequences of one's intended actions as experienced by those most affected by them. Empathy "re-humanizes" the dehumanized enemy. Empathy is at odds with the operant conditioning in infantry training so central to, as Dave Grossman has documented, the post–World War II reduction of the non-firing rate in combat.[6] Empathy runs roughshod over the Marine Corps' "Kill, Kill, Kill" boot camp indoctrination. Often set in the eye—in picturing the impact of one's actions on another—empathy works against war pornography and the whole visual culture regime of killing examined in this book.

This is not to say that the marines in Al Anbar or in the Iraq War overall acted without empathy. The fullest of human endeavors, war cannot exclude compassion from the battlefield—at least not until it excludes human beings as the ones who make it. In the Iraq War, empathic expressions and feelings by marines are scattered throughout the historical record. Before the Third Battalion, Second Regiment crossed the Iraqi border from Kuwait, the Commanding Officer, Lieutenant Brent Dunahoe, told his troops that "civilians should be treated as you would desire your family to be treated in a similar circumstance."[7] Gunnery Sergeant Mike Wynn of the First Reconnaissance Battalion made the same point before the invasion to the marines under him. Having already served in Somalia as a sniper, Wynn implored the younger men to temper the Marine urge to kill: "Don't fucking waste a mother or some kid. Don't fire into a crowd. Those people north of here have been oppressed for years. They're just like us."[8] Once into Iraq, First Reconnaissance sergeant Antonio Espera despaired over the discarded MRE (meals ready to eat) packages and other marine trash strewn on the road in front of several Iraqi homes: "Imagine how we must look to these people." Moments later, a marine relieved himself in open view. Another marine on Espera's fire team comments: "Can you imagine if this was reversed, and some army came into suburbia and was crapping on everyone's front lawns?"[9] Bing West's book on the Iraqi invasion describes marines "agoniz[ing]" over the unintended damage caused by their .50 caliber machine gun rounds; having been told to act with "soldierly compassion," the marines could see how difficult that would be, given their weapons' power and the nature of the operation itself.[10]

A second mode of empathy in wartime can be distinguished as "tactical." The capacity to understand a situation from another person's perspective can be invaluable in battle, especially among senior leaders and commanding officers trying to anticipate an enemy's "next move." Generals, colonels, football coaches, and chess players want to be empathic in this way. Tactical empathy connects back to symmetrical warfare and a ludologist view of battle as a fair fight between equally matched competitors; at the same time, seeing through others' eyes is instrumental to asymmetrical warfare and counterinsurgency's constant need for intelligence. "Intelligence in COIN [counterinsurgency] is about people," the *Counterinsurgency Field Manual* states. "Commanders and planners require insight into cultures, perceptions, values, beliefs, interests, and decision-making processes of the people of the host nation, the insurgents, and the host-nation government." In paragraph 7-8, the *Field Manual* insists that "genuine compassion and empathy for the populace provide an effective weapon against insurgents."[11] Included in the ethics chapter, and under a section on leadership, this provision is one of the *Manual*'s most loaded and confounding. While referencing first mode empathy involving "genuine compassion," it advances that humanitarian sensibility for instrumental value as a "weapon against insurgents." The provision takes one of the warmest and fuzziest of interpersonal dynamics—a cognitive and emotional core of human sociability—and makes it an expedient to warfare. Paragraph 7-8 also assumes discernible differences between populace and insurgents, while other provisions in the field manual stress the likenesses and fluidity between the two. Does sound counterinsurgency doctrine truly merit compassionate empathy among the marines and soldiers doing foot patrols and checkpoints?

The *Field Manual* says little else about ground troops' subjectivity or whether training for counterinsurgency should amend the aggressiveness built into boot camp and infantry school. It doesn't really conceive of marines and soldiers as warfighters. The word "warrior" is used once in the whole text while importantly explaining that counterinsurgency troops may need to accept greater risk to themselves in order to minimize noncombatants' harm: "This risk taking is an essential part of the Warrior Ethos."[12] By comparison, the *Small*

Wars Manual doesn't shy away from the Marines' role as combat troops. A lengthy chapter on training for small wars states that Marine forces "must be thoroughly imbued with the 'spirit of the bayonet'—the desire to close with the enemy and destroy him"; a section on "training en route on board ship" calls for attention to both "street fighting" and "bayonet training" along with talks on the "racial characteristics of country of destination." Another chapter on infantry patrols speaks unabashedly about that operation's primary objective: "to inflict as many casualties upon the enemy as possible. This can be accomplished in a firefight only if the spirit of the offensive, with movement, is employed." At the same time, the *Small Wars Manual* acknowledges that counterinsurgency's complexity may require modifying these imperatives of Marine training. In most small wars, an opening chapter reads, "The aim is not to develop a belligerent spirit in our men but rather one of caution and steadiness."[13] In contrast to the *Counterinsurgency Field Manual*, the older manual contemplates Marine "psychology" for both the violent and "pacification" dimensions of fighting small wars.

The *Small Arms Manual*'s two-pronged approach to the psychology of counterinsurgency informed the effort General Mattis made to recalibrate the First Marine Division for its March 2004 deployment to Al Anbar Province. The belligerent/diplomatic binary can be found in Mattis's well-circulated tenets regarding Marine disposition in Iraq. "No better friend, no worse enemy"—attributable to the first-century B.C. Roman general Sulla—was first uttered publicly by Mattis in March 2003 in one of his many preinvasion speeches to marines amassed on the Kuwait-Iraq border. Then, before the second deployment, Mattis added, "First, do no harm"—a foundational principle in medical ethics. "It was somewhat strange to see a line from the Hippocratic Oath adapted to fit our line of work," Lieutenant Campbell of the 2/4 commented on what he described as the new First Marine Division motto: "First, Do No Harm—No Better Friend, No Worse Enemy." The training that went with that motto, Campbell added—"We crammed Iraqi cultural nuances down [marines'] throats as fast as they could swallow them"—was "very risky, because Marines always have default settings that inform their actions in those precious first moments of a firefight. By training as we did," Campbell concluded, "we flipped our Marines' default switch from 'be fierce' to 'be nice'; we told them to hesitate, to ask questions before shooting, and to assume greater personal risk to better protect the civilians."[14]

Aware of the risk that such a "default setting" for counterinsurgency posed to individual marines and the whole "culture of ferocity" upon which the Corps is built, Mattis carefully balanced every pledge to befriend the Iraqi people with renewed commitment to blowing away anyone and everyone who didn't align themselves with the terms and conditions of occupation.[15] "Be polite, be professional," Mattis told his marines in late 2003 and early 2004, "but have a plan to kill everybody you meet."[16] Such was his extension of Krulak's three-block war and, more specifically, Marine counterinsurgency doctrine into the realities of the postinvasion war in Iraq. Mattis displayed a flair for preceding graphic threats of force with seemingly heartfelt messages of goodwill and peaceful intentions. "I come in peace," Mattis told Shiite sheiks in southern Iraq in spring 2003. "I didn't bring artillery," he added. "But I'm pleading with you, with tears in my eyes: If you fuck with me, I'll kill you all."[17] Mattis's humanitarianism can certainly be found wanting in his no-better-friend-no-worse-enemy approach to counterinsurgency. The same goes, it could be argued, for the general's connection to reality. Friendship? Coming in peace? The first, positive part of each statement is rendered self-evidently false and conditional by what immediately follows.

Given the political context of these utterances—an occupied country and a people having been decimated by shock-and-awe attack and invasion—Mattis may very well have meant to be ironic. That is, Mattis may have realized that Iraqis would not take his statements literally. If they had any say in the matter, Mattis understood, Iraqis would have the heavily armed Marines nowhere near their cities, schools, homes, and families. Clear-headed recognition of this point is in the Marine Corps' history, doctrine, and institutional "warts-and-all" memory of itself.[18] A basic incommensurability exists between the integrity of other people's existence and occupation by a military force—especially by a military force that prides itself, as Mattis described the Marines, as having "an overweening sensitivity to the slaughtering of our enemies." Put simply, the Marine Corps is not a humanitarian institution, although, as discussed later in this chapter, it can be described as an ethical one. In this vein, then, Mattis's communication to the Iraqis worked as shorthand for something like "we can do this either the easy way or the hard way" and "better the velvet glove than the iron fist." Historical record of the Iraq occupation to the Al Anbar Awakening (from 2004 to 2007) reveals the Marines deploying both iron fist and velvet glove. Fallujah I and II in 2004, the killing of Haditha civilians in November 2005, and the continued projection of military force in Al Anbar Province need to be considered alongside the Marines' effort to engage the Sunni sheiks and populace and to restore essential services and infrastructure.

Meanwhile, General Mattis's renown would grow in the direction of a latter-day Teddy Roosevelt or George Patton, as one whose rhetorical gifts have been most memorably used to aggrandize wartime killing. "Mad Dog" Mattis stands as the early twenty-first-century U.S. Marines' supreme ludologist. Before his (in)famous 2005 "It's fun to shoot some people" comment in San Diego, Mattis told two hundred recently deployed marines that same year in Al Asad Base, Al Anbar Province, that "there are some assholes in the world that just need to be shot. There are hunters and there are victims. By your discipline, cunning, obedience, and alertness you will decide if you are a hunter or a victim." He concluded in signature fashion, emphasizing the play element of combat: "It's really a hell of a lot of fun. You're gonna have a blast out there!"[19] Quite essential to the Marine's counterinsurgency doctrine, Mattis meant to vivify in the marines at Al Asad what the *Small Wars Manual* calls "warlike temperament." Having spent generations building a culture of ferocity, the Marines needed to maintain individual killing capacity in the Iraq counterinsurgency alongside any notion of turning off its switch.

Haditha, Acute Stress, and the Excesses of Occupying Force

If the Marines ever worried about counterinsurgency taking the edge off its troops' readiness for combat, the concern would soon be abated by the April and November 2004 Fallujah battles and, overall, by the violent force the I Marine Expeditionary Division applied in Al Anbar Province. To emphasize, the Iraq occupation may very well have been the most difficult post-Vietnam mission for the Marines: clearing and holding Fallujah, Ramadi, Haditha, Barwanah, and other Al Anbar cities of the Sunni insurgency; fighting those insurgents along with the growing number of foreign fighters who came together as Al Qaeda Iraq; and, amid this urban fighting, developing peaceful relations with Iraqi civilians while rebuilding the province's infrastructure. Trying to distinguish noncombatants from insurgents would be a central challenge for Marine riflemen. "Complex COIN [counterinsurgency] operations," the *Counterinsurgency Field Manual* reads, "place the toughest of ethical demands on Soldiers, Marines, and their leaders."[20] This statement, published in 2007, wasn't made in the abstract; it referred implicitly to problems that had already occurred in Iraq. At its worst, the Marines' Al Anbar occupation revealed moral breakdown among troops and officers, with the November 2005 killing of some twenty-four unarmed Iraqi civilians in Haditha the most notorious example. Operating under more restrictive rules of engagement than during the invasion—while IEDs, sniping, and other insurgent tactics effectively evolved—most marines couldn't come close to looking at Iraqis as themselves. Many could not toe that thin line between measured and excessive force. Many could not readily turn off and on the killing switch in keeping with the tactical necessities of the moment.

Senior Marine officers knew the Iraq occupation would be difficult. In a telling interview immediately before the March 2003 Iraq invasion, Lieutenant General James T. Conway, commanding officer of the I Marine Expeditionary Force, stated that he expected Iraqi "bloodletting in the cities" after the United States toppled Saddam Hussein. Because of the sectarian strife that would undoubtedly follow, Conway wanted little part of a postwar occupation: "If I had a vote, I'd say let's get out of here."[21] He preferred that the Marines return to the United States and, in one of the most loaded military terms, "recock" into its ordinary deployment schedules and await the next expeditionary mission.[22] But the U.S. Marine Corps follows orders. And that fact would be underlined in the more official Marine position on the Iraq occupation. "Our Marine Corps has proven to be highly adaptive," is how Lieutenant General James Mattis put it in 2009:

> During Chesty Puller's lifetime [1898–1971], our Corps shifted from what can be described as naval infantry dispersed in ships' detachments, to trench warfare assault troops [World War I], to small war practitioners [Haiti, Nicaragua, et al.], to combined-arms amphibious assault troops [World War II], to extended land operations alongside our comrades in the Army [Korea and Vietnam], to counterinsurgency troops—always maintaining the Corps' ability to defend our country. This adaptation continued in the 1990s as Marines anticipated and prepared for the three-block wars we have been fighting in Iraq and Afghanistan since 2001. Fundamental to our adaptation to today's conflict will be the intelligent initiative of all Marines when the enemy hides among innocent people. This demands a keen understanding of culture—the sort of skill practiced by Chesty and his shipmates in the jungles of Haiti and Nicaragua when they served as advisors to non-U.S. forces. Furthermore, in today's information age, we must recognize that the essential "key terrain" is the will of a host nation's population. This has been demonstrated by our troops in al-Anbar, Iraq, and permits us to gain the trust of

skeptical populations, thus frustrating the enemy's efforts and suffocating their ideology.[23]

The Marines' institutional memory, as Mattis illustrates here, certainly incorporates the most difficult and violent military occupations in U.S. history. Haiti, Nicaragua, among others, involved both efforts to learn the occupied populations' language, culture, and customs—as summarized in the *Small Wars Manual*—and a long, dark litany of atrocities against those same peoples.[24] How does this history fit into the Marines' self-understanding? Ignoring or covering up the details and records of civilian deaths is not beyond the Corps.[25] Haditha proves that. One doesn't have to look very hard, however, for Marines' sober acknowledgment of the realpolitik driving its violence on foreign soil. Retired Major General Smedley Butler's 1930s screeds about the Marines as "gangsters for capitalism" may be the most notorious and incendiary.[26] Nothing less institutionally mainstream than the National Museum of the Marines in Quantico, Virginia, includes a well-placed placard by lead historian Colonel Joseph H. Alexander (ret.) on the Corps' growth in the early twentieth century: "America was suddenly an imperial nation and the Marines had a lot of Uncle Sam's dirty work to do." Iraq, too, was a war of empire. And occupation of that defeated country would be a very loaded proposition. Trying to befriend a people to whom you just laid waste is a deep paradox of the Iraq counterinsurgency. Writing in 2009, after the Al Anbar Awakening ended the worst of the violence in the province, Lieutenant General John F. Kelly hinted at the counterinsurgency's illegitimacy: "In Iraq to a very large degree, we—the U.S. military and civilians—were the source of the insurgency."[27] Incidents of empathy, cultural awareness, and public works projects might help. But from the Sunnis' perspective, there was one abiding truth: the Marines shouldn't be here in the first place.

Within this morally compromised and overly violent historical context, the civilian killings in Haditha seem less an aberration than an unintended yet utterly predictable outcome of the Marines' occupying force amid an insurgency. Haditha—a Sunni stronghold of approximately 100,000 people—had been taken over by the insurgents in 2004 while the Marines concentrated on Fallujah. Public executions of police chiefs and officers and suspected pro-American spies; sophisticated IEDs, with one blast killing fourteen marines in August 2005; uncooperative and shut-down Sunni civilians, as marines patrolled by Humvee empty city streets: "That," as Bing West has written, "was the environment Battalion 3-1 inherited in Haditha in the fall of 2005."[28] On November 19, 2005, a convoy of thirteen marines in four Humvees was patrolling a main Haditha street when a huge bomb obliterated the last vehicle, killing Lance Corporal Miguel Terrazas and seriously wounding two other marines. Shortly thereafter, the Marine squad signaled a car with five Iraqi men to halt; ordered out of the car, the unarmed civilians explained they knew nothing of the blast; one marine recalled later that another marine got down on one knee and shot the Iraqis one by one. After killing the first five civilians, the marines then entered several nearby houses, killing nineteen Iraqi women, children, and old men.

The multifaceted response to the Haditha killings—the media coverage and outrage from Capitol Hill, the official investigations and reports, the legal hearings and trials—rolled out slowly, as the chain of command from the Third Battalion, First Regiment to the Second Marine Division (to which the 3/1 had been temporarily assigned) uniformly decided to ignore the incident. This would be one of the most damning findings when Army lieutenant general Peter Chiarelli, commander of day-to-day military operations in Iraq, finally ordered a full investigation led by Army major general Eldon Bargewell. The Army general's scathing report focused on what it described as the Marine commanders' routine devaluation of Iraqi lives. "All levels of command tended to view civilian casualties, even in significant numbers, as routine and as the natural and intended result of insurgent tactics," Bargewell wrote. "Statements made by the chain of command during interviews for this investigation," Bargewell continued, "taken as a whole, suggest that Iraqi civilian lives are not as important as U.S. lives, their deaths are just the cost of doing business, and that the Marines need to get 'the job done' no matter what it takes."[29] A staggering dissonance can be seen here between Bargewell's documented finding of indifference to civilian deaths and the population-centric mindset that the Marines wanted to instill in its troops going into Al Anbar in 2004.

While Bargewell's report came down hardest on Marine commanders, it could be argued that those officers had quite limited ability to affect the troops' outlook toward Iraqis. In getting at the why of the Haditha killings—and, with that, at the projection of force in Al Anbar—we need to consider the inherent nature of the occupation, any occupation, as a form of domination that brings with it mutual (although not identical) antagonisms between occupier and occupied, a basic hierarchy of the former over the latter. In *Taking Haiti: Military Occupation and the Culture of U.S Imperialism, 1915–1940* (2001), historian Mary A. Renda explains how this "structural" dimension of the Marines' long-term hold on that Caribbean country and its people manifested in a race-based paternalism that then enabled the high levels of violence, including atrocities, against insurgents and noncombatants.[30] While paternalism didn't take on such pronounced forms in Al Anbar Province, race, religion, and the ideology of the global war on terror certainly worked to divide the U.S. troops from the Iraqis.[31] Also, this current study contends that the Marines' capacity for violence had become more automated, located in the eye, and expressed through techniques of killing from a distance—that is, the result of flipping a killing switch. Still, it's worth underlining Renda's point that the structural violence of military occupation diffuses itself into the individual subjectivities of the troops responsible for applying that force.

An elemental difference or dissonance in perspectives comes out in the expressed feelings of those locked in to the Iraqi occupation. "I don't like to say it," as Corporal Timothy Connors, a squad leader in First Battalion, Eighth Regiment, spoke about the occupation in summer 2005, "but after a while, when you have the rifle, and you see how the Iraqis look at you and how they live, then some of our guys feel superior—like the people in Haditha or Fallujah aren't quite human like us." Connors continued self-reflexively: "You don't think of them the same way. That's not right, but it does happen."[32] This "structure of feeling," as cultural critic Raymond Williams described it, would seem insurmountable, as it's set by the very force—"when you have the rifle"—that first defines occupation.[33] And the feeling is mutual. In *The Forever War* (2008), journalist Dexter Filkins recounts conversations he had with Iraqi civilians during the occupation. "I take their money but I hate them," as Filkins quotes Hussein Allawi, a provincial minister for dams: "The Americans are the occupiers. We are trying to evict them."[34] Actually, Filkins explains, the Iraqis carried on two wholly separate conversations: one with and for Americans—"positive and predictable and boring"—made to pacify the occupiers; and the second "among themselves," about "a whole other world, a parallel reality, which sometimes unfolded right next to the Americans, even right in front of them." Filkins adds that when the two worlds did collide, say, in an insurgent attack, Iraqis felt they could not help the Americans. "No one ever saw a thing," Filkins writes in an example that illuminates the anger and frustration the 3/1 squad felt after the Haditha bomb blast killed one and wounded two fellow marines:

> If you were ever lucky enough to get one of the Iraqis to answer, he'd tell you almost without exception that the guys who fired the missile or planted the bomb

or fired from the rooftop had come from somewhere else: outside the village, outside the country. He rarely said, it was that guy, right over there, third house on the left. It wasn't just that the Iraqis lied. Of course they lied. It was that they had more to consider than the Americans were ever willing to give them credit for. The Iraqis had to live in their neighborhoods, after the American soldiers had gone home. The Iraqis had to survive. They had their children to consider. For the Iraqis, life among the Americans often meant living a double life, the one they thought the Americans wanted to see, and the real one they lived when the Americans went home.[35]

Neither the *Small Wars Manual* nor the *Counterinsurgency Field Manual* contains such crystallized insight.

During the worst of the Al Anbar occupation, from early 2004 through 2006, the Marines remained for the most part garrisoned in camp while not clearing, holding, and building. More needs to be discovered about the social history of life in Camp Fallujah and other Marine bases. Reading, movies, music, calls home, and working out took up a good part of the marines' downtime, although computers, with internet access, and video gaming may have been most central. The Marines stocked a "Morale, Welfare and Recreation" center at Camp Fallujah full of Xbox and PlayStation 2 consoles; an Associated Press reporter there in early 2005 wrote that "Marines back from the fight stop in for a few hours to unwind." *Halo 2* (2004) was by far the Marines favorite game at this time. In late 2004, Camp Fallujah's P/X sold its thirty new copies in a few hours. The camp also hosted several *Halo 2* tournaments.[36] Cultural awareness training that the marines received before Al Anbar deployment seems to have dropped away in country. Out of camp, in Al Anbar's city streets, the marines patrolled largely in Humvees—as on the day of the November 2005 Haditha killings. Any notion of friendship or even accessibility also seems to have dissolved. Bing West remembers the marines on patrol at this time looking "like RoboCops, often riding by in armored vehicles with bulletproof windows firmly shut."[37] These were combat troops, whose area of operation was largely hostile. Even if the Sunnis had been so inclined, little opportunity existed for positive meaningful interaction between civilian and rifleman.

A 2006 mental health study of U.S. combat troops in Iraq, including marines in Al Anbar, fills in our sense of the outlook riflemen had toward those under occupation. Authorized by the U.S. Army Medical Command, the Mental Health Advisory Team for Operation Iraqi Freedom conducted its fourth study of deployed troops in fall 2006. The 2006 study and Final Report was the first to include Marines. It was also the first time the team took up battlefield ethics. And that section of the Final Report documents the lack of concern that marines felt toward Sunnis. Of both marines and soldiers, 17 percent agreed that "all non-combatants should be treated as insurgents"—far less than a majority, but still a significant percentage of troops set against a central counterinsurgency tenet. In response to an overarching attitudinal question, less than half of marines (and soldiers) "agreed that non-combatants should be treated with dignity and respect," while "well over a third of all" marines (and soldiers) believed "torture should be allowed to save the life" of a fellow American marine (or soldier). The study's Final Report demonstrates that the marines actual treatment of Iraqi civilians followed their hostile attitudes. In the study 7 percent of marines (and 4 percent of soldiers) reported that they hit or kicked a noncombatant "when it wasn't necessary"; 12 percent of marines (and 9 percent of soldiers) said that they damaged or destroyed Iraqi property "when it wasn't necessary"; and 30 percent of marines (and 28 percent of soldiers) admitted they insulted or cursed at Iraqi noncombatants.[38] If the "no better friend, no worse enemy" mantra still had purchase by late 2006, the marines tilted toward the latter side of the equation, just as they gravitated from the "do no harm" tenet and toward the hunter role of which General Mattis also spoke.

The Mental Health Advisory Team Final Report also divulges the flip side of American Marine hostility toward Iraqis: in large numbers the Marines reported, first, being the targets of violence; and, second, the stress that experience caused. Sixty-five percent of marines (and 62 percent of soldiers) stated that they knew "someone seriously injured or killed or had a member of their team become a casualty." The Final Report authors made special note of how many marines (and soldiers) in interviews spoke of being shot at by snipers; indeed, by 2006 a very popular type of war porn video made and circulated by Iraqi insurgents was of snipers taking out U.S. troops. Perhaps the study's finding most pertinent to assessing the Marines' outlook during the occupation is the significant number of troops who "described an event which occurred during the current deployment that caused them intense fear, helplessness or horror." The Final Report included here short descriptions of "typical" events:

- "My sergeant's leg getting blown off."
- "Friends burned to death, one killed in blast."
- "Mortars coming into your position and not being able to move."

- "A friend was liquefied [sic] in the driver's position on a tank, and I saw everything."
- "A huge fucking bomb blew my friend's head off like 50 meters from me."
- "Marines being buried alive."
- "Working to clean out body parts from a blown up tank."
- "Fear that I might not see my wife again like my fallen comrades."
- "Finding out two of my buddies died and knowing I could do nothing about it."
- "Getting blown up or shot in the head."
- "Just seeing dead people on a lot of missions."
- "I had to pick up my friends off the ground because they had been blown up."
- "My best friend lost his legs in an IED incident."
- "Seeing, smelling, touching dead, blowen [sic] up people."
- "Sniper fire without obvious sorce [sic]."

This material suggests an important explanatory line for the Haditha killings. In asking if one had recently felt "intense fear, helplessness, or horror," the study was testing for *acute stress*—its term for post-traumatic stress disorder (PTSD) symptoms reported by troops in a combat zone.[39] A significant mental health problem, acute stress also closely correlated in the study with marines (and soldiers) who mistreated noncombatants, as it often involves uncontrollable anger as well as shock.

While no mental health study was taken of the 3/1 Marine squad members who survived the roadside bomb in Haditha and who killed the twenty-four civilians, we can speculate for discussion's sake that one or more experienced acute stress in seeing Corporal Terrazas literally blown in two by the bomb—his torso blasted out of the vehicle and landing on the road as his legs remained pinned under the steering wheel. Also, several marines in the squad had fought in Fallujah in November 2004: the Mental Health Advisory Team Final Report emphasizes that those who have frequently faced combat are far more likely to report acute stress; as with its twin disorder PTSD, acute stress accumulates over time, via high amounts and levels of violence in combat.[40] The 3/1 squad members in Haditha that November day in 2005 had been immersed in the Al Anbar occupation about as much any other marines. Their homicidal response to the insurgent attack can be explained in part by the nature of the occupation itself—its considerable length and its express and structural violence.

The violence inherent to the Al Anbar occupation is one explanation why no one in the Marine chain of command ordered an investigation into the Haditha killings. While photographs of the dead civilians passed between marines, the military didn't act on the event until a *Time* magazine reporter started asking questions about it—and after he had seen videotape of the Haditha aftermath taken by an Iraqi journalism student. The 3/1 battalion commanding officer, Lieutenant Colonel Jeff Chessani—who initially responded "my marines are not murderers"—and division commander, Major General Richard Huck—who said "no bells and whistles went off" after he heard about the killings—would both be officially censured for inaction, their military careers effectively ended.[41] Avoiding scrutiny of the killings and the Marines' tactics undoubtedly motivated Chessani and Huck. Although, as seen in interview transcripts from General Bargewell's investigations literally found in a dumpster by the *New York Times*, high-ranking Marine officers explained that the twenty-four dead civilians in Haditha had *not* been unusual for the Al Anbar occupation *after* Fallujah. "Discovering twenty bodies, throats slit, twenty bodies, you know, beheaded, twenty bodies here, twenty bodies there," as Colonel Tom Cariker said in the investigative interview, "you know, we wouldn't see fifteen all the time, but it wasn't—it wasn't something that would be an alarm like, 'Hey, what the hell is going on down there.'" Perhaps the words of Major General Stephen Johnson are most telling about how Haditha was less aberration than representative. Commanding general of the II Marine Expeditionary Force (Forward) in Iraq at the time, Johnson stated that such civilian deaths "happened all the time." The Haditha killings, Johnson concluded, were "just a cost of doing business on that particular engagement."[42]

USMC Literary Culture and Warrior Ethos

What bulwark did the Marines have in place to mitigate the "cost of doing business" in the Al Anbar occupation—that is, to keep more Hadithas from happening during the long war in Iraq? One answer would be the Corps itself: conceiving of itself as a self-correcting ethical institution, the Marines would redouble its effort to instill "Honor, Courage, and Commitment" throughout its ranks. A medium very well suited for such an ideal-driven enterprise is the printed word, and the U.S. Marines could turn to an impressive literary culture for enunciating its "Core Values" and for examining ethical questions arising from what General Mattis described as Iraq's "morally bruising battlefield."[43] The post-Haditha period would see increased attention to—as in the title of a 2012 *Marine Corps Gazette* article—"the ethical warrior and the combat mindset." The warrior archetype reemerged during the late Iraq War as a Marine Corps redoubt, an ancient yet malleable ideal that embodies both the killing prowess and moral fortitude needed in counterinsurgent operations. While instrumental to the projection of military force, the warrior ideal conceives of the warfighter in humanistic terms. The warrior's violence is not governed by a switch to be turned off and on but, rather, is to be applied with honor and with respect for civilian life and, ideally, for one's enemy as well. The "ethical warrior," as described in the *Marine Corps Gazette* essay, is able to override his or her emotions and act mindfully, even when that action involves deadly force.[44] That's a tall order. Literature's strong suit is modeling the fortitude and self-possession needed to pull it off.

Any close discussion of the USMC's literary culture must take in the wide array of its many forms—from manuals on weapons and tactics to graphic novels and pulp fiction, and from popular histories and biographies to introspective memoirs critical of the Marines and U.S. empire. Marine Corps University Press, established in 2008, publishes books and journals on Marine operational history and institutional policy. A Marine literary canon can be identified. Novels include Thomas Boyd, *Through the Wheat* (1923); Leon Uris, *Battle Cry* (1953); James Webb, *Fields of Fire* (1978); and Gustav Hasford, *The Short-Timers* (1979). Memoirs and other nonfictional accounts of one's experience in war are probably the leading form in Marine literature: Elton Mackin, *Suddenly We Didn't Want to Die: A Memoir of a World War One Marine* (1993); William Manchester, *Goodbye, Darkness: A Memoir of the Pacific War* (1979); E. B. Sledge, *With the Old Breed: At Peleliu and Okinawa* (1981); Bing West, *The Village* (1972); Ron Kovic, *Born on the Fourth of July* (1976); Philip Caputo, *A Rumor of War* (1977); and Anthony Swofford, *Jarhead* (2001) are in this part of the canon. A marine wrote each book listed here. Indeed, marines writing about being a marine make up the core of USMC literary culture. In addition to its knuckle-dragging self-caricature, as military historian Max Boot writes, the Marine Corps can be understood as "the most intellectually supple of the services."[45] A certain confidence in books' power to shape minds has contributed significantly to Marine culture and tradition. "Thanks to my reading," as General Mattis makes the point, "I have never been caught flat-footed by any situation." He continues: "It doesn't give me all the answers, but it lights what is often a dark path ahead."[46]

USMC literary culture reaches active-duty marines through the Professional Reading Program. Founded in 1989 by General Commandant Alfred M. Gray, the reading program features the Commandant's Reading List of different titles for officers and enlisted marines (with a few books appearing on both lists). In 2005 General Commandant Michael Hagee revamped the program so that marines could better "develop intellectual adaptability along with effective problem-solving skills." Hagee wanted, as *Leatherneck* magazine explained, to make sure that "Marines get more out of the books they read." The reading program therefore added an interactive website as well as a system for forming and running in-person book discussions among small groups of marines. The program also changed the criteria for books on the Commandant's Reading List.[47] As Marine Corps University president Donald R. Gardner wrote in 2005, "a conscious effort" was made "to raise the intellectual standing of the works by limiting the number of biographies and works of fiction"; rather than being drawn to "fleeting trends," the program chose "serious literature," including "historically based" books that "encompass[ed] broad context[s]" but still "emphasiz[ed] warfighting."[48]

One book that has been on the Commandant's Reading List for many years is *The Village* (1972), by Bing West, a prolific historian of the Iraq War and, earlier, a Marine veteran of the Vietnam War who patrolled with a Combined Action Platoon (CAP) of Marine riflemen and anticommunist militia in Binh Nghia, South Vietnam. Then something of an experiment in fighting Viet Cong, the Combined Action Platoon would become a model unit for counterinsurgency, as the marines lived closely among the Vietnamese and, with that, devoted and sacrificed themselves to protecting the village. The CAP Marines' casualty rate was among the highest of American units in the war. Filled with dramatic accounts of middle-of-the-night patrols, ambushes, and firefights, what makes *The Village* so edifying—what explains its long-standing inclusion on the Commandant's Reading List—is West's careful study of the requisite character and outlook for such a dangerous and complex operation. These twenty-year-old marines wanted to fight. Most were extremely adept at it. But those best suited for counterinsurgency, to put it simply, could also approach and accept Vietnamese people as fellow human beings. West doesn't explicitly include the warrior ideal in his profile of the CAP marine. He doesn't exaggerate the humanity of either the marines or the Vietnamese. It's there though.

And the beginning of the book points out that in starting the CAP program the Marines had great difficulty in finding volunteers who didn't strongly dislike the Vietnamese.[49] Exceptional is the Marine, exceptional is the American, exceptional is the human, *The Village* suggests, who can live among a people different from himself while methodically killing insurgents from that same group of people.

No book has contributed more to the warrior ideal among Marines than Stephen Pressfield's impressive work in historical fiction, *Gates of Fire* (1998), a graphic story of the battle of Thermopylae in 480 B.C. between Persian and Spartan forces (and their Greek allies). While culminating in the Spartan 300's last stand against overwhelming Persian numbers, *Gates of Fire* devotes more pages to the culture that produced such heroism, especially the intense military training program *agoge* and the abiding sense of honor and brotherhood among its warriors. The historical parallels to the U.S. Marine Corps cannot be missed—at least not by marines themselves. As Private First Class Ethan Hoalridge wrote on *Leatherneck.com* in 2007: "The proud history and traditions that make us who we are relate so closely to the Spartans as they were depicted in the book. . . . They were fierce warriors fighting for the brother on their left and right. Marines have many things in common with the Spartan warriors who lived 2500 years ago." *Gates of Fire* is "inspirational and teaches courage and leadership," Hoalridge concluded: "It's on the Commandant's Reading List for a reason."[50] In a type of historical attribution in reverse, Pressfield, himself a marine veteran, drew from Corps tradition to fill in detail of Spartan military culture. Early in the book, Alexandros, a young initiate, one who would be among the Spartan 300 "enunciated in his loudest, clearest voice" to twenty-five hundred warriors before him:

> This is my shield.
> I bear it before me into battle,
> But it is not mine alone.
> It protects my brother on my left.
> It protects my city.
> I will never let my brother
> out of its shadow
> nor my city out of its shelter.
> I will die with my shield before me
> Facing the enemy.[51]

The historical record includes no such Spartan pledge. A "historical invention," Pressfield adapted it from the Marine Rifleman's Creed.[52]

A particular concept that Pressfield emphasized in the warrior ideal is self-control in battle. To be sure, blood and guts abound in *Gates of Fire*. "The ground before the Spartan advance became a sea of limbs and torsos," as one passage describes Thermopylae: "trousered thighs and bellies, the backs of men crawling hand over hand across their fallen comrades, while others, pinned upon their backs, writhed and cried out in their tongue, hands upraised, pleading for quarter." But warrior ethos demanded that a Spartan not lose himself to "blood madness," as Pressfield called it: "The Spartans have a term for that state of mind which must at all costs be shunned in battle. They call it *katalepsis*, possession, meaning that derangement of the senses that comes when terror or anger usurps dominion of the mind." Pressfield's character Dienekes—the Spartan platoon leader who most fully embodies the warrior ideal—had always taught his charges that "war is work." It needed to be "stripp[ed] . . . of its mystery." The book's narrator, Dienekes' squire, observed his master in battle: "His was not, I could see now, the heroism of an Achilles. He was not a superman who waded invulnerably into the slaughter, single-handedly slaying the foe by myriads. He was just a man doing a job. A job whose primary attribute was self-restraint and composure."[53] The construction and maintenance of mental fortitude, over and against what is today called "acute stress"—the terror and anger resulting from combat's murder and mayhem—is, for Pressfield, the basis of the warrior ideal and its ethos.

The presence of mind that Pressfield illustrates in Spartan sensibility would be taken up by early twenty-first-century Marines and applied to fighting the War on Terror. In a 2008 "Ideas and Issues" article in the *Marine Corps Gazette*, Captain Paul L. Croom II draws directly and at length from *Fields of Fire*. "History shows us that in combat, man has always struggled to maintain dominion over his mind. This struggle is not always won, and recent incidents of seemingly blatant and volitional killing of noncombatants by American military forces in Iraq and Afghanistan underscore that katalepsis is a very real problem in today's military, even among the most disciplined." Croom believes that *katalepsis* "is a problem of leadership." Dienekes in *Gates of Fire* demonstrates, for Croom, that "psychological control and command over the physical manifestation of emotions are the hallmarks—and obligation—of a combat leader." Croom zeroes in on the need for Marines to generate and apply controlled force:

As Marines, we train as riflemen. The Corps hones in each of us a killer instinct that capitalizes on aggression and knows no defeat. Without this mindset—and the discipline that instills it—we would not be the elite force-in-readiness that we are but, instead, commonplace among the world's services at arms. When unleashed, however, that rage must be controlled to the brink of the berserker. Absent that control, we degenerate into something subhuman, since it is sapient authority over our actions that separates us from other animals.[54]

By 2008, discussion about the warrior ideal in the Marines had increased considerably. As in Croom's *Marine Corps Gazette* piece, the full-blown self-possessed warrior was offered in counterpoise to civilian deaths and egregious loss of control among some marines in Iraq and Afghanistan. Like Croom and as in the *Counterinsurgency Field Manual*, many characterized warrior ethos as a responsibility of leadership.[55] "To fire their valor when it flagged and rein in their fury when it threatened to take them out of hand" is how Pressfield summarized Dienekes' charge.[56] To take warfighters "to the brink of the berserker," as Croom puts it: leadership in combat, under these terms, is a matter of high-stakes applied psychology.[57] Warrior ethos does not only speak to the demands of leadership, however. A discourse rooted in honor, the warrior ideal could also challenge individual marines to act with purpose, integrity, and commitment.

And that's exactly what Pressfield does in his more recent book *Warrior Ethos* (2011), of which he printed eighteen thousand copies and, all at his own expense, sent to Marines and Special Forces in Iraq and Afghanistan.[58] The book is a short yet ambitious codification of the warrior ideal's ethical components. Pressfield traces the warrior ethos's evolution from its founding in early human history through its crystallization in ancient Greek civilization to its present day formulation in the U.S. military. For Pressfield, the U.S. Marine Corps provides the leading examples of the warrior ethos in practice. As in *Gates of Fire*, while recognizing its origin in war, he distances the warrior ideal in this book from "aggression" and the "will to dominance." The warrior ethos largely grew in principled response to the fear, anger, and darkness engendered by combat. It contextualizes and links self-regard to a larger corporate whole. Camaraderie, brotherhood, and love are its social and emotional manifestations, Pressfield writes:

> There's a well-known gunnery sergeant in the Marine Corps who explains to his young Marines, when they complain about pay, that they get two kinds of salary—a financial salary and a psychological salary. The financial salary is indeed meager.

But the psychological salary? Pride, honor, integrity, the chance to be part of a corps with a history of service, valor, glory; to have friends who would sacrifice their lives for you, as you would for them—and to know that you remain a part of this brotherhood as long as you live. How much is that worth?[59]

To live by the warrior ethos means accepting hardship. "What separates Marines" from other "elite U.S. forces," Pressfield states, "is their capacity for enduring diversity. Marines take a perverse pride in having colder chow, crappier equipment and higher casualty rates than any other service." The warrior ethos also entails "self-restraint." Pressfield wants the reader to remember "the ancient precept that killing is not honorable unless the warrior places himself equally in harm's way—and gives the enemy an equal chance to kill him." While not further applying this principle to modern warfare, Pressfield does emphasize that the ideal warrior's "generosity," expansiveness of spirit, and "capacity for empathy" are needed as much today as ever before. Pressfield means to provoke and inspire in this book. *Warrior Ethos* is to guide the head and heart in combat and in life outside the military. "Let us be, then, warriors of the heart, and enlist in our inner cause the virtues we have acquired through blood and sweat in the sphere of conflict," as Pressfield writes in the book's last sentence: "courage, patience, selflessness, loyalty, fidelity, self-command, respect for elders, love of our comrades (and of the enemy), perseverance, cheerfulness in adversity and a sense of humor, however terse or dark."[60] It's unclear how this creed would have held up amid, say, the worst counterinsurgency operations in Iraq. When set next to combat veterans' accounts of their darkest moments—Kevin Powers writing, for instance, "it felt like there was acid seeping down into your soul"—Pressfield's words sound a bit too sunny.[61] But, it could be said, *Warrior Ethos* does present a toehold if not a bulwark against the excesses of war.

"Which Way Would You Run?"

As the first decade of the twenty-first century came to an end, the Marine Corps had committed itself to producing, in the words of the *Marine Corps Gazette*, "linguistically and culturally enabled marines."[62] As Commandant General James Conway stated in the official planning document *Marine Corps Vision and Strategy, 2025* (2008), "We will go to greater lengths to understand our enemies and the range of cultural, societal, and political factors affecting all with whom we interact."[63] Counterinsurgency operations or "irregular warfare" demanded this cultural turn, an effort led by a team of civilian anthropologists and the 2005 founding of the Center for Advanced Operational and Cultural Learning at the Marine Corps University in Quantico, Virginia. While Marine officers took courses on one of seventeen microregions across the globe, expeditionary units ran through cultural awareness training exercises (in both "live" and simulated programs) before deployment to Iraq or Afghanistan. By 2012, the Marine Corps would showcase in a TV recruitment commercial its capacity for delivering humanitarian aid to people struck by war, famine, or natural disaster. Drawing from its disaster relief operation after the 2010 Haitian earthquake, the Marines' new marketing campaign, "Toward the Sound of Chaos," raised such assistance to a nearly coequal level with warfighting.

After civilian anthropologists settled into Quantico, the Corps gained some distance on the dated *Small Wars Manual*. With the 2006 *Counterinsurgency Field Manual* already pointing out that "biologically, there is no such thing as race among human beings," the Marines' anthropologists revised the *Small Wars Manual*'s view of how race determines national character traits and a people's temperament.[64] In *Operational Culture for the Warfighter: Principles and Applications* (2008), Barak A. Salmoni and Paula Holmes-Eber write that "blanket statements about a cultural mindset failed to capture the important social structures, belief systems, and competing agendas within a cultural group." Having rejected the "culture = mindset = national psychology" equation, Salmoni and Holmes-Eber detail the "five operational culture dimensions" of any "battlespace": physical environment, economy, social structure, political structure, and belief systems. And within the belief systems section, religion receives the most attention. Gone was the *Small Wars Manual*'s description of "uneducated people liv[ing] close to nature" with "childlike characteristics" following "some mystic form of religion"; in its place, *Operational Culture for the Warfighter* offers up early twenty-first-century conventional academic understanding of religion as "locally lived human performances and a locally manifested cluster of beliefs and attitudes."[65] The authors discuss Islam alongside Christianity as one of the world's major organized religions.

In addition to providing culture and language training, the Center for Advanced Operational and Cultural Learning conducts academic-level research on topics related to warfighting. "The Resilience Research Project" (2012), for instance, works from interviews with combat veterans to examine stress and its contribution to the increasing incidents of suicide and PTSD among marines. This ongoing study finds that acute stress arises from dissonance between the

Corps' ideals of honor, courage, and commitment, on the one hand, and, on the other, the frequently grim reality of combat and the self-denigration that goes with it. The "meaning" that a marine attaches to an incident—the report states in using a keyword from anthropological study—is variable and the source of whether one maintains the "flexibility" or "resilience" necessary in the warfighter. Overall, the research project pointed the Corps away from physiological explanations of combat stress and toward greater attention to how wartime violence fits within the marine's life as a whole. "Stress," the study reported, "is existential." To what extent this research will influence Marine Corps doctrine and practice on and off the battlefield is an open question. Senior military journalist Tom Ricks posted a summary of it on his influential blog, and the *Huntington Post* then ran a positive account of the research and how the Marines are rightfully attending to veterans' "crises" in values and meaning that can lead to suicide.[66]

As part of the Corps' cultural turn during the later years of the Iraq War, training and simulation programs concentrated on developing peaceful interactions with noncombatants. A 2010 Marine Corps website article entitled "From Kinetics to Culture" covered a "Realistic Urban Training" exercise at Fort A.P. Hill, Virginia, that featured "village role-players speaking in a foreign language." The article on this "cultural awareness exercise" cites two marines' enthusiasm for the training. While describing the exercise as "a soft knock on the village," Lance Corporal James Chapman seemed quite familiar with what had become the requisite sensibility of counterinsurgency: "You are to be kind and gentle with the villagers. You're just trying to be yourself with them. You want to help them out as soon as possible. They are people who have needs and you want to get it to them because they are suffering." Chapman's company commander, Captain David Bell, elaborated on the idea behind the exercise. "If all we train for is kinetic operations, then we are selling ourselves short and really hindering what the MEU [Marine Expeditionary Unit] was designed to do," he said. "It's an opportunity to dust off Gen. Krulak's three-block war concept," adding that such training still had tactical purpose: "We are here doing the second block by helping the people, meeting with the leadership of the village to find out what their needs are. If we can help them, then we will build a relationship with them and more importantly they will help us. They can tell us where the bad guys are. Then we can go kinetic."[67]

By 2010 the Corps had apparently sharpened counterinsurgency training, just as it had tightened its message about the care Marines would take among noncombatants.[68] A *Washington Post* article covering the Marines' response to the 2010 earthquake in Haiti illustrates how the Corps not only adapted to noncombat operations but also honed its image around such humanitarian aid. Entitled "In Haiti, Empathy Is a Corps Competency," the *Post* article recognizes the huge material assistance the Marines can deliver to people in need: "Long-range heavy lift—the ability to move masses of equipment, supplies and people across the world—is a demonstration of American global influence. It may be the best definition of that influence." The article concentrates, though, on the Marines' efforts at the social level. It reads like a model vignette from the *Counterinsurgency Field Manual*. "As I arrived," the *Post* columnist Michael Gerson writes, "a Marine officer in charge of civilian relations was convening the first meeting of community activists, nongovernmental organizations and local officials. . . . The officer explained to me the complex Haitian class dynamics that determine local leadership. American Marines of Haitian background were speaking a rapid Creole—one of the benefits of a multicultural military. . . . The Marines are practicing a kind of noninvasive surgery," Gerson comments in textbook counterinsurgency form: "providing structure and security, but cultivating community institutions that must continue to stand after America leaves." Writing in an almost astonished tone, Gerson asks "Where in the world did U.S. Marines learn this kind of cultural sensitivity?" He reports:

> Lt. Col. Rob Fulford, in charge of the Carrefour operation, answers: "In Iraq and Afghanistan, where the equivalent was dealing with tribal sheiks. . . . There is a maturity level inside our Marines that didn't exist in 2003 when we invaded Iraq. A cultural awareness. An ability to leverage relationships." Another officer on Fulford's staff adds: "It is very similar to Iraq and Afghanistan, except that here there is no bad guy. We're helping the populace, winning their trust. This is right up our alley. All of us are products of the COIN manual." He is referring to the counterinsurgency field manual authored by Gen. David Petraeus, which involved a dramatic shift in military thinking—increasing the focus on population security, cultivating indigenous capabilities and isolating the enemy by improving the lives of the locals.

Gerson provides here an early historical account of the Iraq War, characterizing counterinsurgency as a success and extending its doctrine to new operations: "This strategy helped save the American mission in Iraq. Its reach and benefits can now be seen on a Haitian beach." One also can't help but imagine how happy the Marines must have been about such positive publicity. Assistance to Haiti involves "a tremendous strain on military equipment," but those involved in the operation "gain skill and experience—moving supplies, treating trauma cases, dealing with complex cultural challenges. Despite many hardships, those who train for a mission love performing their mission," Gerson concludes: "No empire of history could boast such tenderhearted legions."[69]

The notion of tenderhearted Marines comes through in the 2012 multimedia advertising campaign "Toward the Sound of Chaos," led by the J. Walter Thompson Agency, and capped by a television commercial that stands among the most accomplished productions in the Corps' long history of self-promotion. "Towards the Sound of Chaos" included overhauling the official Marine Corps website, which by spring 2013 still highlighted the 2010 Haitian Operation Unified Response, replete with a short documentary on the mission, beginning with the tag line by Commandant General James F. Amos, "We respond to today's crises, with today's marines, today," and ending with the whole campaign's signature provocation, "Which Way Would You Run?" The website also featured the campaign's lead television spot, which first aired in March 2012 during the NCAA men's basketball tournament. The commercial made quite a splash, with both the *New York Times* and *Advertising Age* recognizing its appeal to the "gentler side of service" rather than the Corps' standard branding as the tried and true warrior elite.[70]

Actually, in briefly going back to two earlier Marine television commercials from the post–September 11 era, we catch glimpses at least of the Corps' humanitarian ethos as set within a broader warrior context. "The Climb" (2001), a favorite of early Iraq War marines, hallmarks the Marines' long-standing dare to hardship paired with the promise of brotherhood (fig. 4.1). A young man in pants and t-shirt starts up the face of a rocky jagged cliff, with Marine Corps iconography projected on that surface—the Mount Suribachi flag raising, for example—as he slowly pulls his way to the top. A marine in World War II fatigues extends a hand to the recruit as he finishes the climb. "The passage is intense," the voiceover declares, "but if you complete your journey, you will find your destiny among the world's greatest warriors." The recruit is now a marine, standing tall at the very top in dress blues: "The few, the proud, the Marines." One-third of the way through the commercial, though, on the cliff face, right before the flag raising, we see a moving picture clip of a marine feeding a girl from his tin rations cup.[71] The same imagery appears in the television commercial "Pride of the Nation" (2005), a montage of photographs

Counterinsurgency and "Turning Off the Killing Switch" 123

Figure 4.1. The 2001 U.S. Marine Corps commercial "The Climb" features the individual challenge of becoming a marine and the brotherhood that awaits "the few, the proud, the Marines." Aired on network TV during weekend sporting events, especially NFL football games, the commercial was commonly mentioned by marines deployed in Iraq as having caught their attention before enlisting. (United States Marine Corps)

chronicling Corps history, including one of the Marines hitting the beach during World War II. The humanitarian image appears halfway through, as the voiceover heavily emphasizes the word "compassion."[72]

Developed after the Iraq War, and with the Afghanistan War winding down, the Corps' "Toward the Sounds of Chaos" advertising campaign looked beyond the Global War on Terror and envisioned the Marines as the nation's first responders to crisis abroad. The "Which Way Would You Run?" commercial rehearses that role, illustrating the Marines' mobility and firepower, while pledging its capacity for humanitarian assistance (fig. 4.2). The spot opens to the distant sound of machine gun fire and women screaming for help from a stark, ominous, smoke-filled desertscape (seconds 1–5); the shaky-camera point-of-view shot continues with close ups of military boots, legs, and then marines running frantically "towards the sounds of chaos" (seconds 6–10). It's shot brilliantly, with strobe effect accentuating the tautness of the marines' back and leg muscles: the moment couldn't be more desperate, the response couldn't

Figure 4.2. The 2012 U.S. Marines commercial "Which Way Would You Run?" is part of the Corps' "Towards the Sound of Chaos" multimedia effort to promote itself as a humanitarian relief force. While demographic research suggests heightened interest in service among young adult Americans, the notion of marines engaging in humanitarian aid is hardly new. It is featured, for instance, in General Charles Krulak's influential 1999 essay on "strategic corporal" and "the three block war." (United States Marine Corps)

be more strenuous, the Marines' commitment couldn't be more intent—one can see it on their faces as they run forward to the unknown. The exact content of that response appears at the 60-second commercial's very center: transport trucks carry large boxes stamped with an American flag and clearly marked "Aid" (seconds 29–31). "Ready to respond at a moment's notice," the voiceover states, "they are the first to move towards the sound of tyranny, injustice, despair." Finally, as the words "Which Way Would You Run?" appear white on black at the end of the spot (seconds 59–60), the commercial revises the familiar Marine dare into a double dare: do you have the courage to run toward chaos and, in the most positive light, do you have the courage to do so for the sake of helping rather than shooting others?

Lest anyone doubt the Corps' commitment to warfighting, the "Which Way Would You Run?" spot does show off the Marines expanded capacity for projecting force in the second decade of the twenty-first century. It effectively promotes the highly distinctive Marine Corps brand of mobility and firepower delivered from the sea. While the commercial's geographic location is nondescript, the operation clearly happens from the sea. A signature moment of the commercial comes with the marines literally hitting the beach: after a few frames of dark screen, the door of a landing vehicle drops open, kicking up sand, with blue ocean in the background and Marine infantry pouring out and heading inland (seconds 19–24). The Marines are "ready to respond, at a moment's notice" (seconds 12–15), because it has the U.S. Navy to take them pretty much wherever they need or want to be. With the help of the Navy's aircraft carriers, along with the country's airbases throughout the world, the Marine Corps relies heavily on its own air power. In fact, the recruitment commercial showcases MAGTF: Marine Air Ground Task Force, the Corps' optimal unit for expeditionary warfare. The 2012 commercial exhibits every attack aircraft type in the Corps: including F-18 and Harrier fighter jets, Cobra helicopters, and V-22 Osprey. And Marine ground forces are deployed from and across the sea, in this commercial at least, by helicopter as well as by hovercraft and amphibious landing vehicles.

As the Marines aired "Which Way Would You Run?" on national television, the head of the Corps' recruiting command, Brigadier General Joseph L. Osterman, spoke about how the commercial responded to a generational interest in humanitarian assistance. No, that doesn't mean the Marines are "getting soft," Osterman insisted. "There is a subset of millennials," though, as Osterman explained, "who believe that the military is an avenue of service to others. Not only in our nation but also in others faced with tyranny and injustice." He was referring to a national online survey the J. Walter Thompson Agency conducted in developing the "Toward the Sounds of Chaos" campaign for the Marines. It found that many young U.S. adults consider "helping people in need, wherever they may live" to be an important component of good citizenship.[73]

A demographic trend toward service may help to hasten a transformation in the U.S. military that General Osterman wasn't ready or able to acknowledge. The Iraq and Afghanistan Wars may be the last significant ground operations relying on predominantly human forces. As the Pentagon pours billions of dollars into unpiloted aerial vehicles, ground robots, and other weapons systems for posthuman warfighting, and as the psychiatric and medical professions discover more about the close links in the Iraq and Afghanistan Wars between the horror of combat, face-to-face killing, post-traumatic stress, and suicide, wouldn't it be good to think that U.S. marines and soldiers will take on a more exclusive role as warrior protectors with less violence visited upon them and the nation's enemies. The problem with that vision is that it would require the United States to decide to project less force across the globe. Otherwise, the nation will make war (if that's what it will be called) by increasingly automated means, which may very well increase rather than lessen the level of death and destruction overall.

CHAPTER FIVE

Posthuman Warfighting

Is it possible that the U.S. Marine Corps *is* "getting soft"? While its commercial branding still features mobility and firepower—the capacity to project massive force inland from the sea—few would deny that the Corps faces major transformations in the early twenty-first century. Developing cultural sensitivity in an infantry force that has long taken pride in killing people and breaking things may be the least of the Marines' challenges, although the presence in Quantico of practitioners of arguably the most humanist of academic disciplines—anthropology—is a new element certainly worth watching. The imperative of sexual equality, which includes ending sexual violence, is the most overdue transformation. Many marines tend to view women in combat as a sharp stick in the eye of its hypermasculine grunt (read infantry) culture, but, as we saw in the previous chapter, the warrior ideal is already being rewritten in more gender-neutral terms.

The Corps' interservice identity in the post–Iraq War U.S. military has undoubtedly attracted the greatest attention among Marine leadership. Secretary of Defense Robert M. Gates told a Marine audience in 2010 that the country no longer needs a "second land army," a role the Corps took on in Iraq and Afghanistan. He also expressed doubt over the continuing feasibility of its "large-scale amphibious assault landing" signature: "New anti-ship missiles with long range and high accuracy may make it necessary to debark from ships 25, 40 or 60 or more miles at sea."[1] High-tech weaponry has forced the Corps to adapt or die, Lieutenant Colonel Lloyd Freeman wrote in his widely circulated 2013 *Foreign Policy* article "Can the Marines Survive?" Picking up on Gates's speech, Freeman asked pointedly: "Why does the Marine Corps insist on maintaining such a large amphibious entry capability based around the same Marine who stormed ashore at Tarawa?" Even more fundamentally, Freeman questioned the need for an infantry-based service in an era when "land forces will no longer win wars." Instead, Freeman concluded, "Computers, missiles, planes, and drones will."[2]

The rise of unmanned aerial and ground vehicles for combat as well as for intelligence, surveillance, and reconnaissance (ISR) may be the most overarching challenge to the Marines in the early twenty-first century.[3] Warfighting with drones and robots means less danger and fewer casualties for U.S. troops. But is that really what the Marines want? Priding itself on leaving no marine behind, that value of care for "our own" arose within the context of high-risk expeditionary operations that featured the corollary value of honor through sacrifice for country, Corps, and comrade. What happens to the Marines' brand if risk taking, and the promise of adventure and heroism that goes with it, are lost in wars waged via automated weapons systems? Some critics of reliance on high-tech weaponry worry that it devalues humanity and lessens hope for humane war making. They're skeptical that the warrior ethos—along with value judgment, conscience, and even empathy—can or *should* be coded into a combat robot's artificial intelligence.

A related host of issues stems from the cybernetic fusion of warfighter and machine. The combat robot is not the archetypal goal in this enterprise but rather it's the superhuman warrior whose strength, endurance, mobility, and other physical capacities are augmented by exoskeleton technology, metabolic engineering, and pharmacological enhancements, and whose perception and cognition are implanted with synthetic optics, digital information technology, and data processing. Warfighters with "no physical, physiological, or cognitive limitations will be key to survival and operational dominance in the future" is how one official from the Defense Advanced Research Projects Agency (DARPA) put it in 2005.[4] DARPA has been following this prosthetic impulse for some time. There's reason to believe now, though, that, on the heels of the Pentagon's huge expenditures on research and development for the Iraq and Afghanistan wars, technological innovation is reaching a new critical mass— one that can be called a "posthuman" age of war making.

It shouldn't be surprising, then, that this chapter begins with science fiction. For decades science fiction writers and filmmakers have explored the military future through marines. If not always the U.S. Marines, it's been "space marines" or "colonial marines" who have carried that expeditionary spirit and mastery of small-arms fire into intergalactic adventure and force projection. At the same time, the Marine Corps has looked to science fiction for ideas about doctrine, weapons, tactics, and training. The fact that a particular weapon system or other technology "exists in our movies proves that it is potentially possible," USMC colonel James Lasswell (retired) said in 2006. "If you can imagine it," he continued, "we think it can happen."[5] The "we" Lasswell referred to are those at the Marine Corps Warfighting Laboratory in Quantico, Virginia, where his team once developed a research and development request for jet bikes or "Jedi Broomsticks" (from *Star Wars: Return of the Jedi* [1983]) to help marines avoid IEDs in Iraq.[6]

We've already seen an example of science fiction influencing military development in chapter 3's discussion of the Pentagon trying to blur the line between actual combat and video gaming. The master text here is Orson Scott Card's 1985 novel *Ender's Game*, one of the most successful works in science fiction, and a book that has long appeared on the Marine Commandant's Reading List. What the Corps wants its readers to get from *Ender's Game* is an open matter though. Effective tactics through close teamwork is one theme that the Marines would undoubtedly support. Ender's improvisational skill on the battlefield may be another. The book raises the issue of whether one is morally responsible for killing—in Ender's case committing what Card calls "xenocide," the killing of another species—when the warfighter believes he or she is acting within a simulation and not against real live enemy targets. It's a timely question, although perhaps one the Marine Corps isn't interested in exploring very closely.

In addition, *Ender's Game* examines the relationship between empathy and war. Ender's military genius is based largely on his ability to feel what his enemy is feeling. While invaluable in battle, empathy can also be construed as war's opposite. Indeed, that may very well be Card's point. The book can be read just as much as a critique as an endorsement of simulation training and digitized warfare. Ender hadn't intended on killing the alien species Buggers. Wracked with guilt, he now dedicates his life to learning everything about the Buggers and, against his superiors' wishes, telling their story as they saw it. "I stole their future from them," Ender explains: "I can only begin to repay by seeing what I can learn from their past." What he learns, and what he tells the world, is that strength comes not from "battling" but through cooperation and "love." For the Buggers, "It was the beginning of wisdom."[7] Ender spreads this message to other planets being colonized by humans as he becomes Speaker for the Dead, one asked to come to know the life of a deceased person and to relate it from the distinct perspective of the one who lived it.

Over time, Speaker for the Dead evolves into its own worldview, a type of humanist religion whose sole principle is the importance of one's will, motive, and intention. In referring to the adult military supervisors who had him kill the Buggers, Ender says, "It did not occur to them that this twelve-year-old boy might be as gifted at peace as he was at war."[8] While surveying a distant planet, Ender comes upon the alive pupa of a Bugger hive queen, already fertilized by male larva, cocooned away for protection, and apparently awaiting salvation by the one who killed every other member of her species. Ender is now a redemptive figure. The novel ends with him carefully taking the cocoon and setting off on a journey for the right time and place for the queen to awaken and give birth back to the Buggers.

The conclusion of *Ender's Game* brings us back to the question of whether the Marine Corps is getting soft. Choosing peace over war, creating a religion based on deep knowledge of and empathic feeling for others: Ender stands in sharp contrast to, say, the iconic Marine drill sergeant or even the intellectual Marine officer who recognizes the historic need for military force and for training people to kill; the story's expansive treatment of the species' capacity for compassion runs against the Corps' essence of assault warfare. Maybe marines don't pay much attention to the end of the book. Or, conversely, maybe humanitarianism really is gaining some ground in the Corps.

The U.S. Marine Corps has a Martial Arts Program, founded in 1999, which combines physical training with "character discipline" and ethics training. Two marines closely involved with developing the program, Jack E. Hoban and Joseph Shusko, explained in a 2012 *Marine Corps Gazette* article that the "combat mindset" of the "ethical warrior" is one of "protector" rather than "killer." This is no small distinction, the authors recognized: "Killing only to protect life" doesn't neatly fit with the Marine's history as colonial shock troops. They asked if "protectors [are] as prepared for the realities of war as killers? Will ethics training somehow make us soft and less capable of accomplishing the mission?" While insisting that a protector can be fierce and dangerous—think of a "grizzly bear with two cubs"—Hoban and Shusko conceded that they "don't know the answer definitively." They ended the article, though, by promoting the "dual life value" of the Marine Warrior: "Our respect for the dual life value of self and others safeguards our humanity and sets us apart from our enemies who do not respect the lives of others outside of their 'in group.' "[9]

A principled and practical ideal, dual life value fits with anthropology's primary ethic of respecting the integrity of all other peoples' existence, while it may also mitigate post-traumatic stress (PTS), since "research has shown," as Hoban and Shusko wrote, that "dehumanization of the enemy exacerbates PTS."[10] Whether it works with U.S. foreign policy may, again, be the sticking point. But as we proceed to discuss the Marine Corps in an era when warfighting may very well be delegated to drones and robots, it's worthwhile to pay attention to evolving humanitarian values in the U.S. military. Something of a commonplace in military history is that warfighters are shaped by the weapons they're given for killing the enemy. If that killing power is now outsourced to nonhuman entities—autonomous artificially intelligent machines—do humans in the military grow closer in nature to those automatons or is it the opposite? With the taking of life technologically assured through the mediated perception of the enemy target—the killing itself a potentially mechanical action of stimulus and response—might humans populating the Marines and other military services distinguish themselves from those machines and grow in their humanity by redoubling protection over unnecessary violence?

Marines in Science Fiction and in Space

Gene Roddenberry didn't include marines in *Star Trek* (1966–69). Produced amid mounting opposition to the Vietnam War, the original series is humanitarian in its intentions, setting the starship *Enterprise* on largely peacekeeping missions—although its overarching mission, "to boldly go where no man has gone before," betrays the TV show's Cold War origins and expansionist ethos. Captain James T. Kirk loves a fight. And the *Enterprise*'s capacity for projecting force is considerable: its antimatter Transporter beams landing parties into "strange new worlds," armed with phasers invariably set to stun; and its antimatter photon torpedoes usually keep the Klingons and other enemies at bay. But missing from the *Enterprise*, its crew members, and from the culture of the twenty-third-century Federation Starfleet are a distinct warrior sensibility and any semblance of marines.

Otherwise, the post–World War II burst of science fiction in literature, as well as in TV, movies, and video gaming, is filled with marines. A few of these texts are made in such close cooperation with the U.S. Marines that they work as product placements or recruitment commercials for the Corps. The Hollywood film *Battle Los Angeles* (2011) is one such work, with its alien-fighting plotline featuring CH-46 Sea Knight helicopters and hard-nosed grunt warfare in equal measure. Far more science fiction representations of marines are generic. With plot lines built on adventure in space, they often include small contingents of heavily armed marines responsible for on-board security and military expeditions on new planets. Consider the *Warhammer 40,000* franchise (1987–present). Starting out as a tabletop game, the British-owned *War-*

hammer now includes books, films, and video games, all of which feature Space Marines, chosen among the best warriors in humanity—with aggression and killer instinct their defining characteristics—and augmented with genetically modified organ implants. *Warhammer*'s Space Marines operate at the Imperium's frontier. With their primary role a rapid-strike planetary assault force, they are nothing if not mobile, equipped with a host of vehicles, including Jetbikes, Thunderhawk Transporters, and Landspeeder Tempests.[11]

Similar themes, equipment, and vehicles can be found in U.S.-produced science fiction works. In the movie *Aliens* (1986), for instance, heroine Ellen Ripley travels back to a hostile planet with heavily armed and armored U.S. Colonial Marines. In 2013 the videogame *Aliens: Colonial Marines* continued the *Aliens* franchise. Played from the perspective of United States Colonial Marine Corps (USCMC) Corporal Christopher Winter—a strategic corporal in space—the first-person shooter's wiki explains that USCMC is

> the successor to the United States Marine Corps and is the United Americas' primary "force-in-readiness," who are at all times ready for immediate deployment without any additional reinforcement, training or provisioning. They specialize in force projection, being able to operate independently in environments far from home for extended periods thanks to their technological prowess and sizeable space fleet at their disposal.[12]

Colonial Marines also appear in the cult TV series *Battlestar Galactica* (2004–9). Part of the Colonial Fleet, the black-helmeted marines secure the ship from Cylon incursions. In the season two episode "Fragged," Colonel Saul Tigh, furiously battling the Cylons on a nearby planet, dramatically orders, in a vivid example of product placement, "Send in the Marines!"[13]

In *Avatar* (2009)—the all-time highest-grossing film worldwide—marines are central to the story set in the mid-twenty-second century. Its hero, Jake Sully, is a disabled veteran Reconnaissance Marine, who becomes mobile again through his avatar on the planet Pandora, which humans are mining for its natural resources while a private security force (made up of ex-marines) tries to counter an insurgency by the indigenous population. Something of a controversy developed over what some took to be writer and director James Cameron's less than positive depiction of marines in *Avatar*. Cameron sharply contested the point, saying on the *Jay Leno Show* that his brother was a U.S. Marine—"I got nothing but respect for those guys"—and explaining further that in making the movie he spoke with marines deployed in Iraq and Afghanistan about counterinsurgency operations.[14] In a different vein, in *Avatar*'s climactic fight scene, Sully dons an Amplified Mobility Platform—an exoskeleton—whose hydraulics give him Herculean strength, enough to carry two giant 30 mm auto-cannons into battle. A design supervisor for the studio responsible for the movie's special effects called the exoskeleton "an Apache helicopter with legs."[15]

Among the countless militaristic-themed works in science fiction, one of the most successful and influential is *Starship Troopers* (1959), Robert A. Heinlein's novel about the twenty-second century Mobile Infantry and its war against the alien Bugs. Heinlein's "starship troopers" are not marines, although his book's ties to the U.S. Marine Corps have long been noted. Heinlein himself graduated from the U.S. Naval Academy. Very familiar with Marine Corps culture, he used it in *Starship Troopers* to animate the Mobile Infantry's "fighting spirit." For instance, chapter 1's epigraph is taken from Sergeant Daniel Daly's call to his marines to attack the Germans at Belleau Woods in 1918: "Come on you apes. You wanna live forever?"[16] Like the U.S. Marines, Heinlein's Mobile Infantry is transported to combat by the Navy. *Starship Troopers* has been on the Commandant's Reading List for decades. The book "raises many issues," as one Marine reviewer put it, "that deserve the consideration of all those involved with preparing people to undertake violent action on behalf of their governments."[17] In addition to *Starship Troopers*' detailed description of Mobile Infantry training and the discipline that goes with it, the book introduces various technological innovations in warfighting to which the Marines and U.S. military have paid close attention.

The most attention has been given to the Mobile Infantry's "powered armor," an exoskeleton far more advanced than the one worn by Sully in *Avatar*. Technically speaking, Sully's should be called a "mecha"—a large robotic machine operated by a human who sits in a cabin inside the system. Heinlein's powered armor, by contrast, is specially fitted to the trooper's body. It moves with the wearer's limbs. "The real genius in the design," as Johnny Rico, the story's lead trooper explains, "is that you *don't* have to control the suit; you just wear it, like your clothes, like your skin."[18] In addition to carrying a wide array of weaponry—some hold small nuclear warheads—the two-thousand-pound suit of powered armor greatly augments the trooper's strength, speed, senses, and jumping ability. The suit's legs have jet boosters that allow the wearer to jump several stories high and keep the trooper, in one of the Mobile Infantry's favorite phrases, "on the bounce."

There's a reason why *Starship Troopers* has been influential for so long. Heinlein was prescient. Like Orson Scott Card recognizing in *Ender's Game* the effect that electronic simulation and video culture could have on military training and reflexive violence, Heinlein envisioned in *Starship Troopers* how the infantry could remain a centerpiece of warfighting by augmenting the marine or soldier's sensory and physical capacities. The trooper's powered armor includes sophisticated mapping and communications systems that link him visually and verbally to others in the platoon. The suit also has a pharmacological apparatus used to control, among other things, the trooper's energy level during battle.

For the U.S. Marines, perhaps the timeliest aspect of *Starship Troopers* is how Heinlein made the infantry mobile in space. While powered armor keeps the troopers bouncing on planet surfaces, they get to the fight on spaceship troop carriers and are then dropped from orbit via individual reentry capsules. Heinlein offered considerable detail about the capsule system: "Precisely like cartridges feeding into the chamber of an old-style automatic weapon," the troopers are shot out of the spaceship's "twin launching tubes." It's the only time that the battle-hardened Rico "gets the shakes": waiting for your turn to go, "you sit there in total darkness, wrapped like a mummy against the acceleration, barely able to breathe."[19] Layers of the capsule burn off upon entering a planet's atmosphere, as "chutes" slow the trooper's descent until his suit's jets help set him down. "Cap troopers" is another name for the Mobile Infantry, which, in Heinlein's work of science fiction, is truly a space-borne infantry.

Back to the early twenty-first century, the Marine Corps recognizes that amphibious assault via beachheads will no longer be its hallmark operation among the U.S. armed forces. The point became clearer in 2011 when the Pentagon scrapped development of the amphibious Expeditionary Fighting Vehicle, a $15 billion program designed to replace the Assault Amphibious Vehicle, or "amtrack," which the Marines have been using since 1972. Even though an enemy's long-range guided missiles remain a threat to U.S. Navy carriers, the Marines' plan for inserting infantry for combat or emergency relief, as of spring 2013, is still to launch from Navy ships bobbing some twenty nautical miles offshore but to land further inland, close to the operation's target. "We don't see fighting for beachheads anymore," as Major General Kenneth McKenzie Jr. explains: "We'll enter where the enemy isn't, or suppress parts of his defenses and reach deep inland." Crucial to this doctrine of direct force projection is air power. The Marine Corps is looking forward to 2017, when the new F-35B Lightning II fighter jet will be deployed and used for taking out enemy anti-air defenses; unlike the F/A-18, which is currently assigned such missions, the F-35B takes off vertically and therefore doesn't rely on aircraft carriers. Also central to this version of expeditionary warfare is the MV-22 Osprey tiltrotor aircraft, which can transport both lightweight howitzers and up to thirty-two troops. The concept behind this new expeditionary warfare is called "distributed operations": rely on small platoon-size infantry units that can be inserted quickly and, if need be, make time for a larger joint force.[20]

Within this revised doctrine of expeditionary warfare, there are room and reason for Space Marines—that is, Marine Corps infantry transported by way of suborbital spaceships that could drop a platoon anywhere in the world within two hours of being ordered to do so. Neither the U.S. Marines Corps nor the Pentagon is quite ready to match science fiction in this regard; no one from the military has spoken on the record about developing a *Starship Troopers*–like capsule system for deployment; but since 2002 there has been a small group of Marine Corps officers working on SUSTAIN, the Small Unit Space Transport and Insertion. In 2006 the Marine Corps worked with *Popular Science* magazine to imagine what such force projection would look like, and in 2010 the Pentagon produced a Concept of Operations document that sets out how SUSTAIN would be used. An initial reason for SUSTAIN came about during the beginning of the Afghanistan War, when it proved exceedingly difficult to gain permission from Pakistan and other countries to fly through their air space. "What if we don't have to have anybody's permission?" then lieutenant colonel Roosevelt Lafontant asked: "What if we could just go above and drop in?" Once the Marines land and complete the mission, extraction would become another significant issue.[21] But, within the constant quest to shorten the timeline for deploying American forces, Space Marines may be an ultimate answer—at least until DARPA produces *Star Trek*'s antimatter Transporter.

Another key intersection between science fiction and military planning and development is the exoskeleton, which some in the Marines see as a practical answer to the mounting dilemma of infantry combat load. Writing in the *Marine Corps Gazette* in 2005, Chief Warrant Officer Jeffrey L. Eby describes the absurd amount of "great [new] combat gear" that he now has to carry: a global positioning system; special rifle sights for the M-16; a helmet with night vision goggles and otherwise "adorned with more gadgetry than my family Christmas tree"; up to two or three radios; a K-bar and Gerber multitool; two canteens and a Camelbak personal hydration system; an improved first aid kit; a butt pack; Hatch protective gloves and a wrist compass; knee pads; two ink pens and notebook; binoculars; two map pens; a whistle and flashlight; a spare knife; and an Iridium phone. "All of this is before I receive my mission essential ammunition load," Eby adds. "The assault load of the average infantryman," he continues, "drain[s] the energy of the fighting Marine before contact with the enemy is made." In response to this problem, he turns to *Starship Troopers* and Heinlein's "strength-enhancing exoskeleton suit," as Eby describes it: while protecting the trooper from "chemical or biological worries," it "cooled him during hot weather and warmed him during cool weather. It allowed the wearer to jump over buildings and run across countrysides faster than today's vehicles. There were built in communications, optics, protection, and strength. One individual attacked and destroyed entire cities that today would take corps-level forces if not a field army."[22] As fantastical as this sounds in the early twenty-first century, the Pentagon is pushing ahead with a military exoskeleton project, albeit a system initially without the firepower or bounce capabilities of the powered armor found in *Starship Troopers*.

The Marines may find exoskeletons effective in rebuilding the Corps' infantry capability. As Elby argues, exoskeletons could alleviate the combat load problem and thus would increase the all-important mobility of Marine infantry for maneuver warfare. "The exoskeleton is a radically new technology," Major C. Travis Reese writes in his 2010 master's thesis for Marine Corps School of Advanced Warfighting, "that offers ground forces the potential to influence the fight persistently, over great range, and more diverse terrain than ever imagined." The "system would be a progression in force employment concepts for the future," Reese concludes, "that would see foot mobile forces operating at a tempo nearly equivalent to aerial and vehicle-borne forces."[23] Finally, highly trained infantry fully outfitted in an exoskeleton replete with armor, weaponry, communications, optical augmentation and other cybernetic implants may compare favorably to combat robots or "battlebots," which pose an ultimate challenge to the centrality of humans on the battlefield.

The Postmasculinist Marines and New Optics of Combat

Among the far-reaching questions springing from technological advancements in U.S. war making, at least one goes directly to gender politics and the unfolding integration of women into combat roles. Would marines and soldiers fighting in exoskeletons keep their gender? That is, if the time comes when infantry wear powered armor that augments the wearer's lethality and other physical and perceptual capacities—augments them in a uniform manner, bringing each warfighter up to the same levels—then what meaningful gender differences remain between those men and women while in the suits? The question isn't fanciful. The trajectory of twenty-first-century military technology intersects with the quest for gender equality within the armed forces. Like the exoskeleton, most new weapon systems and accompanying battlefield technologies result in the *disembodiment* of infantry in combat. As proximity with the enemy and brute force continue to recede from the act of killing in war, secondary sexual characteristics—such as strength, foot speed, and endurance—will lose importance in identifying those best suited for the infantry. This is a familiar argument for women in combat, but here I'm suggesting something more. With exoskeleton enhancements, remote-controlled robots, and long-distance killing via high-tech optics, a new cognitive and affective paradigm is emerging in ground warfare: further and further removed from the age-old masculinist traits of red-in-tooth-and-claw ferocity and pugnacity, it's a sensibility determined and characterized by genderless cybernetics, a combat technique located far more in the targeting eye than the overpowering body.[24]

With the Pentagon's January 2013 decision to end the ban on women serving in combat units, many Marines fear that the Corps' warfighting standards, traditions, and culture will be greatly compromised. "To say that this issue has lit up . . . the Corps is an understatement," observed John Keenan, U.S. Marine Corps colonel (retired), and editor in chief of the *Marine Corps Gazette*. Arguments abound for maintaining the ban.[25] The most standard are that women generally lack the physical capabilities necessary for combat and that, even if some women have those capabilities, their inclusion will hurt unit cohesion. Retired U.S. Marine Corps general Robert H. Barrow's emotional testimony before the Senate Armed Services Committee on Women in Combat in 1991 is still cited as the authoritative statement on why women shouldn't be placed in combat roles. Before declaring that their inclusion "would destroy the Marine Corps," Barrow responded to the point that modern war no longer has frontlines and that therefore women in support roles are already in danger:

> Combat is a lot more than getting shot at. Or even getting killed. . . . Combat is finding and closing with and killing or capturing the enemy. It's killing, that's what it is. . . . Brutality. Death. Dying. It's uncivilized and women can't do it nor

should they even be thought of doing it. The requirements of strength and endurance render them unable to do it. And I may be old fashioned but I think the very nature of women disqualifies them from doing it. Women give life, sustain life, nurture life, they don't take it.[26]

A moralistic tone can be heard in Barrow's incredulity over women "even being thought of doing" such "uncivilized" work. First lieutenants B. L. Brewster and R. K. Wallace expanded this moral argument more than twenty years later, writing that "'women and children first' has been a tenet of all emergency rescue efforts since time immemorial, yet we are evidently now prepared to dismantle this fundamental principle of western society for the sake of equality." Theirs is a heartfelt essay, an appeal to male chivalry the likes of which seldom appears in today's public discourse. Entitled "Let Us Fight for You," the 2013 article by Brewster and Wallace desperately urges the Marines to maintain traditional gender roles. "Men are to shoulder the responsibility of fighting to preserve the life and dignity of women, as well as to protect the next generation," the authors state as a type of imperative. And, more anxiously they ask: if women are allowed in combat, "Where now does a man go to prove his manhood in society?"[27]

In addition to such moral reasons for maintaining men as the only marines in ground combat, current arguments for excluding women also invoke human nature, a biological force that precedes chivalry or any other cultural value or practice. Brewster and Wallace include this analytical line in their brief against women infantry. "From an evolutionary perspective," they write, women in combat would mean "pitting a generally smaller/gentler/more compassionate demographic against a generally larger/stronger/more violent demographic in a 'survival of the fittest' contest that ultimately determines the fates of societies."[28] In rhetorical terms, to cast something as natural, rooted in biology, is to make it immutable and predominant; any measure outside of that dynamic— such as weak and compassionate women in battle—is wrong and doomed to fail. The biological argument is most prevalent within the unit cohesion discussion. Sex drive, especially the human male sex drive, is a natural force. Placing women in close-quarter infantry units would trigger that drive, which, in turn, would only keep the marines from their military purpose. In a follow up to General Barrow's statement before the Senate Armed Services Committee, U.S. Marine Corps colonel Mackubin T. Owens wrote in 1992 that "the behavior of men around women," unlike race relations (in an allusion to the racial integration of the military), is not "learned" and cannot be "changed." More recently, Brewster and Wallace put it more straightforwardly: "No matter how disciplined the man, sexual attraction—or at least distraction—is as involuntary as a woman's natural response to a baby's cry."[29] The male marine would either act on his feelings or expend considerable energy in subsuming them. Either way, the unit is no longer a focused "band of brothers," as Barrow put it. The unit is no longer cohesive.

These marines have a point. Sexuality is (at least partially) biologically determined. Although to say that Marine training and discipline—a regime of laws, rules, and practices beginning in boot camp—could not shape intraunit sexual behavior and even feelings is difficult to accept. This is the same military service that takes diverse young recruits and conditions them to become warfighters, a process that includes, by the Corps' own reasoning, overcoming instincts for self-preservation and strictures against taking another human life. Omitted from contemporary theorizing about how women are not natural-born killers is the fact that the current consensus on the psychology of combat holds that the human species as a whole is not given to killing in battle. The anti-female-infantry commentators may be correct, then, in arguing that putting women in combat is unnatural—just as it is for men.

In another important sense, these marines may be right in saying that female warfighting runs against human nature. If human nature is understood to be the result of biological *and* cultural evolution—the long comprehensive arc of the species' history shaped by genetics as well as by choice, contingency, and drift—then it's not incorrect to conclude that men have been the "natural" warfighters. Exceptions in myth (e.g., the Amazons) and history (Soviet pilots and snipers) of women in war can be found, of course. But, beginning perhaps with early divisions of human labor based on body size, reproductive capacity, and expendability, war making has by and large been the province of men. As some feminists would have it, war can even be seen as a male invention, the dark and destructive product of masculine instinct, proclivity, and desire.

Within this macrohistorical framework of war, it's possible to characterize the next stage of military development as "posthuman." If the nature of pre-twenty-first-century warfare stemmed from initial physical differences between men and women, and if the technological innovations of today and the near future take killing in combat so far beyond what has been contained in and achieved by the human body and its physical force, then we might label this next stage *posthuman warfighting*. The term is most applicable in relation to projecting force via, as discussed in the following section, unmanned attack vehicles, including combat robots.

At the least, we can envision in the not-too-distant future a *postmasculinist* era of the U.S. military and the Marines. Women will continue to move into combat roles. A crucial point that's gone missing from the male-only position is that huge social and cultural changes regarding sex roles and gender relations have *already* taken place in the United States since World War II.[30] In contrast to the racial integration of the military in the post–Korean War era, when the armed forces helped lead the transformations occurring in American society at large, the Pentagon is behind in eliminating inequality based on sex and the sexual violence that helps support it. The deeply rooted values in U.S. political culture of individual autonomy and merit-based opportunity dictate that the military follow what has already happened in education, business, the professions, and sports. Only women able to meet the infantry's physical standards will be eligible for combat, but that option must be open to them. In a highly militarized society such as that of the United States, fighting for one's country carries with it honor, civic status, and opportunities in politics, business, and other public realms like no other type of job or service. Accumulation of social and political capital is the most powerful reason for ending the ban on women in combat.

Coincident with the pressure to include women in the infantry is what we've already seen as the possible "softening up" of the Marines in the winding down of the Iraq and Afghanistan wars. Conceiving of marines as protectors— those who kill only to save the lives of others—is one ethic that may grow in this new era. If so, like the grizzly bear protecting its cubs, the role of protector would fit well with preexistent ideals of womanhood. Author and retired marine Stephen Pressfield recently discussed the inclusion of women within the warrior ethos. "The lioness hunts. The alpha female defends the wolf pack," Pressfield writes: "The Warrior Ethos is not, at bottom, a manifestation of male aggression or of the masculine will to dominance." He continues: "Its foundation is society wide. It rests on the will and resolve of mothers and wives and daughters—and, in no few instances, of female warriors as well—to defend their children, their home soil and the values of their culture."[31] The hypermasculinity that has accompanied war making for generations is far from extinct. To be sure, with the smallest percentage of women among the U.S. armed forces, the Marine Corps remains a testosterone-driven bastion of traditional martial culture. There's reason to believe, however, that even the Marine Corps is changing. Books and moving pictures about the Marines fighting in the post–September 11 wars, while filled with the pornography of violence, are not fixed upon masculinity as much as their counterparts from the first half of the twentieth century. As this study argues, early twenty-first-century warfighting has been motivated less by physical prowess idealized in the male warrior form and more by acquiring techniques of violence located in the eye and expressed through skills of killing from a distance.

To emphasize, the geographic distancing of the killer from the killed also distances combat from the embodied hypermasculinity long associated with war. If an exoskeleton works to dissolve gender difference among its wearers, so too do weapons systems that function from a great distance, which are triggered by push buttons and touch screens, and which rely on advanced optics. Current research and development in optical systems that couples perception with artificial intelligence will further separate warfighting from pugnacity, aggressiveness, and other gendered emotions of the battlefield. For instance, DARPA is working on a Cognitive Technology Threat Warning System that employs 120-megapixel cameras to monitor a battle space; the recorded images are then fed into multiple computers running complex cognitive algorithms designed to detect enemy threats—what's called a "cybernetic hivemind"; finally, the results from the computations will be fed into smartglasses featuring Head Up Display—picture Google glasses for the warfighter.[32] The system's net effect will be to project into a corner of the warfighter's field of vision the totality of the cameras' data that is already categorized or interpreted for potential hostile action. Linked into the "cybernetic hivemind," the warfighter will act in response to an optical intelligence collective, taking infantry training's see/shoot imperative to a far greater order of magnitude and thereby bypassing individual human perception and the emotional field that has gone with it.

Finally, the Marines' battlefield dictate in the Iraq War of "getting eyes on it" will entail systems that will not only see farther but will also perceive in different ways. The *2012 U.S. Marine Corps Science & Technology Strategic Plan* elaborates on how small-unit "distributed operations" will depend on ever more sophisticated optics. "Distributed units" will use "lightweight, organic, manned and unmanned platforms and advanced weapons for the rapid, accurate, effective application of firepower across the full range of military operations," the *Plan* states. To locate enemy forces, "Marines will employ integrated, lightweight optics and sensors to see through all battlefield conditions (day, night, low light, and obscuration)," an allusion to new research and development that will "see" through or around solid objects such as walls of buildings.

Detection may be a more accurate word than *sight* in describing the perception involved in some systems under development. With the Marines' goal to further "observe, classify, track, record, and report information on enemy movements, habits, and intentions," the *Plan* calls for "small, light-weight, autonomous sensors that can capture data with respect to the electromagnetic spectrum, acoustic, seismic, and magnetic properties."[33] Future counterinsurgency operations will warrant crowd-scanning systems utilizing face recognition optics along with such features as detecting thermal blood flow patterns that signal aggressiveness and other markers of hostile intention. Per *Starship Troopers*, such sensors could also be fitted to combat helmets and smartglasses, or even implanted within the warfighter's head itself. A science fiction commonplace, the cyborg warrior is being slowly constructed in today's military, where its extrahuman sensor systems will contribute to the dissolution of gendered sensibility among male and female marines and soldiers.

The Gladiator Robot and the Critique of Remote Warfare

The most significant development in U.S. war making over the past decade has been the transformation of unmanned aerial vehicles (UAVs) from largely an intelligence, surveillance, and reconnaissance (ISR) role to a hunter-killer role. Pursuing and taking out Taliban and Al Qaeda fighters in Afghanistan, Pakistan, and Yemen from video screen and joystick consoles in Nevada is at once effective in its deadliness and controversial for a host of reasons, including its propensity for killing the wrong people, the extralegal omnipotence it bestows upon the commander in chief, and the risk-free nature of the killing for the "pilots." Those piloting the missions report a wide range of feelings and thoughts about their job. On one end of the spectrum, there's the infamous declaration that "it's like a video game. It can get a little bloodthirsty. But it's fucking cool." Other drone pilots have spoken more gravely about flying over Afghanistan with a deadly machine one moment and driving home to American suburbia the next. Some reportedly suffer from "high operational stress," "clinical distress," and PTSD.[34] Within U.S. military argot, the unmanned aerial vehicle is a "force multiplier" for both its ISR and hunting-killing capabilities. Meanwhile, critics argue that the drone has crossed a remote-killing threshold that places it outside of warfare and the rules, practices, and laws governing and defining war for centuries.

The Marine Corps doesn't directly use drones. Or at least it hasn't up until the ends of the Iraq and Afghanistan wars. Unmanned ground vehicles (UGV) have been added to the Corps' forces, however, including the Dragon Runner—a small, remote-controlled reconnaissance-scout robot used since 2007 to "see around the corner" in urban combat and other environments. It is designed "to be the eyes and the ears of the Marines in forward urban operations," its chief architect explains, "allowing them to gather information without being in harm's way."[35] Light and durable enough to be hurled over fences and walls, Dragon Runner's controller system, as its project officer Captain Dave Moreau describes, is quite easy to use: it looks, feels, and works like "a portable video game."[36] The Marines also have bigger things in mind when it comes to UGVs. So-called PackBots have already been "deployed" for transporting the hundreds of pounds of supplies and equipment for an infantry unit. The biggest idea within military robotics is "robot warriors," systems free of direct human control that will take on clearing operations and other high-risk aggressive missions. What the Marines—as an institution and as individuals within the Corps—think about the prospect of being replaced by "battlebots" is a complex question, one whose answers will only unfold over the next years and decades.[37] In one vein, it's difficult to imagine such technology not being developed and, with that, it not being implemented. "The trend is clear," says Ronald Arkin, a leading robotics engineer at Georgia Institute of Technology: "Warfare will continue and autonomous robots will ultimately be deployed in its conduct."[38]

The Marine Corps took a step in the direction of warrior robots in 2003, when—at the beginning of the Iraq War and amid the growing threat of improvised explosive devices (IEDs)—it partnered with Carnegie Mellon University to produce a prototype of what has been described as "the world's first multipurpose combat robot" (fig. 5.1). According to the Carnegie Mellon National Robotics Engineering Center's description, the Marines wanted the golf-cart-sized Gladiator not only to support infantry units but also to be out in front of them. The Gladiator "gives infantry a way to remotely perform combat tasks," Carnegie Mellon's project description reads, "which reduces risk and neutralizes threat. It is designed to support dismounted infantry in missions that span the range of military operations," while the Gladiator's "operator and supported unit can remain concealed when it goes into action, improving their safety." As for what it carries into action, the Gladiator prototype is loaded for bear: a machine gun with six hundred rounds of ammunition, a multipurpose assault weapon with antitank capability, a light vehicle obscuration smoke system, and an antipersonnel obstacle breaching system.[39] With such firepower and

Figure 5.1. 2007 prototype of Gladiator Robot, built by Carnegie Mellon University's National Robotics Engineering Center for the United States Marine Corps. While the Marines decided in 2007 not to develop the Gladiator prototype at that time, remote-controlled and autonomous robots for combat will likely emerge as a key component of U.S. force projection in the near future. (© Copyright 2015 National Robotics Engineering Center, Carnegie Mellon University)

capacity for leveling obstacles, the Gladiator certainly would have been appreciated by the Marines in Fallujah among other urban battles in Iraq.

Like the Dragon Runner, the Marines' Gladiator prototype is operated with a video game–like console and display. Feeding into that display is a sophisticated optics system that "perform[s] as well," claims Marine Corps lieutenant Colonel (retired) James McMains of the Office of Naval Research, "as an individual Marine with currently fielded binoculars and thermal imaging equipment." That says a lot. In addition to a marine operating the Gladiator at considerable remove, the warrior robot's own optics outdistance those of most if not all opposing infantry units. Day and night video cameras, an integrated position warning system, a laser rangefinder, and an acoustic detection system may be just as crucial to the Gladiator's battlefield effectiveness as the robot's weapons systems.[40]

After Carnegie Mellon built the Gladiator prototype in 2007, the Marines decided against pursuing the program further. With the anticipated winding down of the Iraq War and budget cuts likely to follow, the Marines may have thought that it simply needed to invest its resources elsewhere. Further discussion about the Corps' reasoning would be speculation, since there appears to be no document in the public record explaining its decision. But the Marines' rejection of such a powerful automated weapon system built to protect infantry from danger begs the question of whether a bottleloot is in keeping with or in opposition to the warrior ethos that is so central to the Corps' institutional identity and self-understanding. In one sense, the Corps wants to protect its own—it has a moral responsibility to do so. In another sense, though, to reduce or possibly eliminate risk from killing in combat threatens values of bravery and courage while also raising complex ethical issues over the projection of such asymmetrical posthuman force.

The prospect of developing and deploying a warrior robot force pretty much dissolves any notion of war as a blood sport fought between two equally matched opponents within mutually recognized conventions of what constitutes a fair fight. True, this ludologist conception of war is inaccurate and dubious: human warfare, from the beginning, has constituted the effort to maximize threat to one's enemy while minimizing risk to oneself; and the aggrandizement of armed conflict as a game and via a sense of play can easily be rendered anathema amid the mass destruction of total war. Yet, as we've seen in chapter 3, ludology continues to hold sway in the Marines. From recruitment and training through video gaming to motivational speeches to troops going into battle, the idea and ideal of combat as competition—a test of one's bravery, cunning, and prowess—remain at the center of the U.S. Marine Corps' enterprise and identity. General James Mattis's wartime quip that "it's fun to shoot some people" is only the simplest iteration of the ludologist sensibility.

Warrior robot engineers, project directors, and other proponents have argued for replacing human troops with automated and autonomous weapon systems. Writing in the *Marine Corps Gazette* in 2008, U.S. Army major John Bohlin insists that, "regarding conventional arms, robots are the next, and potentially greatest, force multiplier since the advent of artillery, the repeating rifle, the machineguns, or air power, respectively." He goes on to say that even though deploying a large force of "combat ready robots" (say, 40,000 to 50,000) is oftentimes considered to be "impractical and astronomically expensive," the reality of such an investment in the "long term" would be a "reduced drain on the national economy and greater productivity from the military's human resources." The cost of training and supporting a soldier will be "astronomically higher" than that of a robot once its "assembly is streamlined."[41]

Another advantage warrior robots would have over infantry is their freedom from human hesitation to kill one's own species. As discussed in the introduction, post–World War II U.S. military training has worked from the assumption that human beings are instinctively resistant to killing in battle. Applied behavioral psychology (in the form of see-shoot training) alongside an increasingly lurid visual culture of war pornography, as summarized in chapter 2, reduced the nonfiring rate from supposedly 80 percent in World War II to 5 percent in the Vietnam War. With the deployment of warrior robots, nonfiring rates would no longer be an issue. Battlebots "will be stone-cold killers," says weapons analyst John Pike, "and they will be infinitely brave."[42] That very point, though, leads to what may be the biggest drawback to developing and using warrior robots: fear of relinquishing human control over killing in war.

With presumably this concern in mind, the Defense Department commissioned a study by Georgia Institute of Technology roboticist Ronald Arkin to determine if autonomous robots could be made to follow ethical values that would guide their killing in battle. Yes, Arkin has answered somewhat unsurprisingly (given his earlier investment in such a proposition), warrior robots could be bestowed with what he calls "artificial conscience"—a behavioral framework built, in part, with pertinent international laws of war and specific rules of engagement. What may be surprising is that Arkin has taken it a huge step further, saying that warrior robots would "ultimately behave in a more humane manner" than human warfighters. Without feeling "fear, anger, frustration or revenge," Arkin writes in his technical report, *Governing Lethal Behavior in Autonomous Robots* (2009), battlebots would indeed be "stone-cold killers." With no empathy blocking the way to taking human life, they would also be free of the emotional trigger to noncombatant casualties. Robots would act more humanely in battle, Arkin adds, because, without concern for their own mortality, they wouldn't follow the human impulse to protect oneself in uncertain situations by firing at unknown (potentially noncombatant) targets. While warrior robotic optics and other sensory devices are not near the sophistication needed for such discriminatory killing, Arkin argues that it is only a matter of time before an autonomous weapon system's perception would far outdistance that of humans and thereby allow it to better distinguish enemy fighters from civilians in low-visibility environments.[43]

The prospect of developing autonomous robots superior to humans in both killing capacity and humanitarianism, as it were, has sparked considerable criticism of the whole posthuman warfighting enterprise. British international relations scholar Christopher Coker argues that war is an intrinsically human institution; and for military leaders "to subcontract responsibility for our actions to robots" is to distance the species from its humanity. Even if robots' so-called ethical judgment in high-risk operations may be trusted more than that of human beings, Coker suggests, the outsourcing of war's violence will have long-term repercussions on human existence and moral constitution.[44] Killing by remote control or with autonomous robots, for instance, may very well lower to a dangerous degree the threshold for going to war. War would be cheap. With little risk to lives in the nation that uses such weapons, leaders would find war politically inexpensive and would therefore be less motivated to pursue alternatives to it.

U.S. development of weapons systems based on remote and automated killing marks a new inequality in warfare that transcends asymmetry, rendering enemy troops harmless to American forces and, with that, removing the ethical basis for taking another life in war. The potential for warrior robots fully equipped with operational moral consciences doesn't get around this ethical problem. Just the opposite: by substituting human presence on the battlefield with a machine (whether autonomous or controlled remotely), the key justification for killing in war—the mutual threat soldiers pose to each other—is not just compromised but, again, eliminated. "At the heart" of this problem, as legal scholar Paul W. Kahn explains, "is a violation of the fundamental principle that establishes the internal morality of warfare: self-defense within conditions of reciprocal imposition of risk. Without the imposition of mutual risk, warfare is not war at all." What exactly is it then? As Kahn answers, "it most resembles police enforcement," a function the Marine Corps and Army will likely have to take up more and more, even though that role isn't in closest keeping with the warrior ideal and identity.[45]

The moral basis for attacking enemy combatants dissolves with this shift from projecting force via warfare to doing so via international policing. "The moral condition of policing," Kahn explains, "is that only the morally guilty should suffer physical injury." But combatants, by virtue of that identity alone, "cannot be equated with the morally guilty," Kahn continues, "since opposing combatants are likely to have equally valid claims to moral innocence. Neither has wronged the other, or anyone else." Attacking the regime under which the combatants serve may be morally justified, but, to emphasize, the combatants themselves, once they pose no real threat, can no longer be distinguished from noncombatants as appropriate targets. Within this framework, the targeting with remote or automated weaponry of nonstate actors who have already committed violent acts may be justified because, it can be argued, they are morally guilty—a distinction consistent with dual-life theory's principle of killing only to protect life. But otherwise, a state capable of killing others "with no human intervention but only the operation of the ultimate high tech weapon," as Kahn puts it, is a state acting without any "sources of constraint" and is therefore "an intolerable political situation" to other states and to the populous whose leadership wields too much power to be fully trusted.[46]

In conclusion, over and above the ethical critique of remote weaponry and warrior robots stands the issue of what consequences such push-button killing have on the humanity and even the humanitarianism that has heretofore accompanied war making. The robotocist dream of a "moral governor" in killing machines misses the point that morality is situational, contextual, and embodied: it's a practice and action within human perception and judgment, not code to be uploaded. Even if warrior robots could be developed to more closely follow certain rules of engagement, the removal of moral agency from human warfighters places organized killing of human beings on a new and, by definition, inhuman plane of force projection—no matter how humanitarian the Marine Corps and other U.S. armed forces may become in the twenty-first century. It's a difficult problem. Since humans invented war, armies have tried to limit unnecessary risks to their own forces; military leaders, it can be argued, are morally bound to do so.[47] And yet those same military leaders may be among the first to oppose discarding the warrior ethos that recognizes the nobility in placing one's mortal body at risk to protect others.

CHAPTER SIX

Synthetic Visions of War: Conclusion and Epilogue

It was July 4, 2012, in Quantico, Virginia. I was doing research for this book at the Marine Corps History Division. Because of the holiday, the archives were closed, so I went next door to the National Museum of the United States Marine Corps—something I very much wanted to do anyway. It's an impressive military museum. Layered in historical depth, modern and interactive, the museum bursts with pride, knowledge, and celebration of its subject. It couldn't be otherwise.

Making my way to the World War I gallery, I came to perhaps the museum's most popular feature, an exhibit of one of the Marines' proudest moments: Belleau Wood, the battle fought by the Fourth Marine Brigade in France against advancing German troops in spring 1918, a time when the Corps did as much as any other to substantiate itself, in its own accounting and others, as a fierce and formidable fighting force. Belleau Wood produced some of the Corps' most canonical lines and attributions: Captain Lloyd Williams's reflexive response to an anxious French officer's advice, "Retreat Hell! We just got here"; Sergeant Dan Daly's moto, "Come on, you sonsofbitches! Do you want to live forever?"; and "devil dogs," the sobriquet the German command supposedly bestowed upon the Marine Brigade after the battle.[1]

With such a cornerstone of Marine memory, myth, and tradition, the National Museum curators decided to represent Belleau Wood in motion picture. Completed in 2010, the exhibit takes up a whole wing of the museum and includes a walkway through an early twentieth-century cityscape with audiotape of a newsboy spouting headlines from the war. One then steps into a room with its sidewalls painted to resemble the French forest. On the far wall is an eight-foot tall and fifteen-foot wide high-definition screen accompanied by twelve speakers surrounding the visitor. Playing in a three-minute loop is a high-production-value reenactment of a Marine charge across an open field of tall grass against well-protected German machine gun placements. With Steven Spielberg–like realism of whistling bullets and point-of-view shots from both

American and German perspectives, the Belleau Wood exhibit is something altogether different from the rest of the museum's display cases of weapons and uniforms, historic set pieces of camp and front, and touch-screen features of famous figures and battles.

Upon entering the Belleau Wood room, the film already playing, I noticed four young men clustered together, their faces aglow from the lit screen, and their eyes transfixed by what they saw there. With high and tight haircuts and wearing Marine fatigues, they were all in Quantico's Officer Candidate School. I had seen their bus parked in the lot. No better way to celebrate their country's birthday. Without the nerve to ask, I could only imagine what they were thinking and feeling while watching the film. They certainly weren't smiling. There was tension in their visage and posture. And I started to understand why as I turned to the screen myself.

There may be a heroic undertone to the Belleau Wood film. All the onscreen marines—some actors, some real grunts from Quantico—have ridiculously sharp and jutting chins. They walk steadfastly across the field, slightly crouched against the machine gun fire, but leaning in unwaveringly. Most of them are killed, though, before they get anywhere near the enemy. The Marines don't pull that many punches—at least not here. I watched the film through several times. A few marines do break through the German line, but there's no victory or even resolution to take in before it ends. The film mostly shows marines getting hit by German fire. It is historically accurate: the Fourth Brigade lost more than forty-seven hundred men, more than half of its strength, in the three weeks of fighting at Belleau Wood.[2] But for the National Museum to portray the death so plainly is quite unusual within popular visual culture of war. I go back to those four young men, some of whom may be in Afghanistan today. At the same time, some may have walked away that day and—having seen war a bit more closely for what it is—may have just kept walking, away from the Corps, the U.S. military, and foreign wars altogether.

Not likely, though. An assumption behind this study is that the marines tend to know their work. And here the Marine Museum curators have well understood their audience and how to instill in the several hundred thousand American civilians who visit the museum each year a visceral sense of appreciation for Marine courage and sacrifice. The steely resolve embodied in the Belleau Wood film certainly fills that bill. As for appealing to Marine recruits—like the four officer candidates I saw that Fourth of July—the film says something a bit different: it's less a matter of "respect how others have died for your country" and more "this could be you" and, at least in this one instance, "you better see and understand what you're headed toward." There's long been a type of death wish attached to leading renderings of American bravery. "Swift pictures of himself, apart, yet in himself, came to him" is one way Stephen Crane's *The Red Badge of Courage* (1895) envisions the death instinct contained in war: "a blue desperate figure leading lurid charges with one knee forward and a broken blade high—a blue, determined figure standing before a crimson and steel assault, getting calmly killed on a high place before the eyes of all." Or, as Marine veteran E. B. Sledge put it flatly in his celebrated World War II memoir, *With the Old Breed: At Peleliu and Okinawa* (1981): "Slowly the reality of it formed in my mind: we were expendable." He may not have actually wished for his expendability or have been driven by its completion, but, as Sledge's account makes so clear, his bravery in fighting the Japanese in the Pacific occurred in accordance with accepting the fact of his very likely death in battle. In detailing the intergenerational presence and the Corps' tradition that animated his combat experience—"with the old breed"—Sledge also speaks of the "stark realism" the Marines have cultivated toward killing and death in battle. Within this context, the Belleau Wood film is nothing more or less than a well-placed piece within a much larger whole.[3] It suggests that Marine Corps' culture and history bestow a certain cognizance of the meat grinder effect in modern warfare.

Biopolitics and the Costs of War

The National Museum's larger than life enactment of Marine sacrifice at Belleau Wood is part of a vast visual culture of combat and war that, as this study makes clear, not only reflects and illustrates the Marine Corps' fire power but constitutes its capacity for projecting military force itself. My interest in the optics of combat was sparked some time ago by John Keegan's elemental work, *The Face of Battle* (1976), and its historical focus on the existential question, "What is it like to be in a battle?" Keegan's reconstruction of the foot soldier's "personal angle of vision" at Agincourt (1415), Waterloo (1815), and The Somme (1916) spurred my research into *what* and *how* the U.S. Marines *saw* during the 2003 Iraq invasion and the consequent occupation of that nation and its people. I grew particularly concerned with the equation between sight and the high levels of violence expressed in both the invasion and the occupation. The disparity between World War I troops' "very limited view" amid a "cloud of unknowing," as Keegan described, and the early twenty-first-century

Marines' (technologically and otherwise derived) *clarity of vision* only sharpened my critical interest in how the contemporary American eye for battle has equaled U.S. proclivity for projecting force overseas.[4]

Compared to the what and the how of this subject matter, the *why* of combat optics and its relation to U.S. empire isn't as easily construed. Claims of causation, responsibility, and blame run to several presidential administrations, myriad public and private institutions, technologies, and long-term processes of directed development, historical drift, and contingent change. Military historian Andrew J. Bacevich's book *The New American Militarism: How Americans Are Seduced by War* (2005) points to "several disparate groups" that shared the goal of moving the country past the defeatism of the immediate post–Vietnam War period. "Military officers intent on rehabilitating their profession," foreign policy intellectuals fearing weakness abroad, religious leaders worried about the loss of traditional values, "politicians on the make," and "purveyors of popular culture"—"as early as 1980," Bacevich writes, each forwarded a type of American militarism that creates "a romanticized view of soldiers, a tendency to see military power as the truest measure of national greatness, and outsized expectations regarding the efficacy of force."[5] From Bacevich's study I argue that U.S. militarism further manifested itself in presenting ways (for soldiers and civilians) to see the world, including one's self and country and its potential enemies, a sensibility that represents combat and military operations in general as clean, effective, and gratifying. By focusing on the U.S. Marines and the open-ended myriad institutions interconnected with the Corps, I have sought to uncover and determine the *structures of knowledge, feeling,* and *perception* that helped to produce such a long-term and massive act of violence as the Iraq War.

In my treatment of combat optics and other aspects of militarized thought, feeling, and perception, one finds a pronounced aversion to the apparent ease with which the United States began and carried out the 2003 attack and occupation of Iraq. Citing the numbers of dead and wounded Iraqis and Americans helps betray the scale and asymmetry of U.S. military power. My study tries to connect the early twenty-first-century eye for battle—as best exemplified in the Marines—to what might be called "biopolitics": critical attention to the ways that systems of power affect human bodies, persons, and flesh at a basic species level.[6] Understanding the optics of combat as such a power system gets at the Iraq War's mournful differential of destruction, the high magnitude of premeditated violence that stemmed from, as Indian novelist Arundhati Roy predicted in her essay "Algebra of Infinite Justice" (2001), the systematic devaluing of non-American lives in response to the September 11 attacks. "How many dead Iraqi women and children," Roy asked searingly and presciently, "for every dead American investment banker?"[7] Biopolitics also goes to a second tragedy of the Iraq War: the wholesale nature of bodily, neurological, social, and psychological harm done to the American men and women who served in the conflict. Instrumental in motivating individuals to enlist, fight, and kill, the optics of combat examined here are abjectly incapable of mediating the human destruction involved in war.

A primary factor in the early twenty-first-century projection of U.S. military power has been the *all-volunteer force*, established in the 1970s after the Vietnam War, and the ensuing separation of public opinion, and of the American citizenry overall, from Washington's (the Presidency plus the Pentagon's) proclivity for making foreign war. To be sure, steadfast reliance on ever-higher tech weapon systems has contributed to the free exercise of military force. Dead and wounded American troops, even if they volunteered, are an increasingly serious political liability for a wartime president. As discussed in chapter 5, distancing the killer from the killed through remote technology has been effective in strengthening force protection—limiting U.S. losses—and thereby better assuring the continuation of a particular military action. Having a well-paid, highly trained professional force that will, as Bacevich puts it, "go anywhere without question," has also enabled the commander in chief, especially in the first stages of a new military deployment.[8]

With the Iraq and Afghanistan wars being the first extended military campaigns in American history to use an all-volunteer force, historians will want to learn about the experiences of the highly trained and skilled soldiers and marines who proved to be both so effective in combat and so delimited in other aspects of those conflicts. The U.S. Marines will undoubtedly continue to attract a large share of historical attention. If there's a warrior class in the United States, it runs directly through the Marine Corps, and it exemplifies a proclivity for killing that, among many recruits, may very well lay just beneath their person's surface, with boot camp being just as much about learning how to control violence as it is about unleashing something that doesn't come naturally to them. As we saw in chapter 4, the Marines spoke in Iraq of turning off and on the "killing switch." For those historians interested in developing a biopolitical perspective on the Marines' killing culture, one issue may be how to treat the troops who enthusiastically take on the rigors of war: does that somehow mitigate concern over the systemic harm that comes to *them* in combat?

If biopolitics concentrates on how power systems reach in and organize, control, and curtail human bodies within war, then it would also have to address the *knowingness* that many young American men and now women exhibit with regard to the killing and danger that will be part of their service in the infantry. There's an assumption of risk, a pleasure seeking, and a this-is-part-of-my-job sense of sacrifice that need to be factored into understanding how militarism and war work upon human subjects. "What unites them is an almost reckless desire to test themselves in the most extreme circumstances," is how journalist Evan Wright sized up the marines in the First Reconnaissance Battalion during the 2003 Iraq invasion. His book *Generation Kill: Devil Dogs, Iceman, Captain America and the New Face of American War* (2004) captures the marines' self-consciousness:

> In many respects the life they have chosen is a complete rejection of the hyped, consumerist American dream as it is dished out in reality TV shows and pop-song lyrics. They've chosen asceticism over consumption. Instead of celebrating their individualism, they've subjugated theirs to the collective will of an institution. Their highest aspiration is self-sacrifice over self-preservation.

In historical terms, it's quite possible to see the Marine Corps having made up one elaborate dare to young Americans' manhood, a dare to accept the physical challenge and mortal danger of combat. If we recognize that Marine recruitment has often connected most with those who perceive that they have the least to lose in their civilian lives—"these young men represent what is more or less America's first generation of disposable children" is how Wright summed it up—then a biopolitical view of war would do well to consider that modern U.S. society retains a sacrificial ethos, an impulse that incorporates a seat-of-the-pants cost-benefit analysis of military service generated by the country's gross inequalities of income and education, racial hierarchies, and other material differences among its citizenry.[9] There's also a crude evolutionary logic to sacrificing young men. With human sperm far less scarce and valuable than eggs, as sociobiologists theorize, young men, especially of the poorly educated and impoverished kind, are a nation's most expendable demographic. Interestingly, though, this same sacrificial dynamic can help make sense of women starting to move into combat roles through the infantry. Modern U.S. feminism has fought to distance womanhood from sexual reproduction. Bodily strong and fit young women may pick up the Marine Corps dare of the infantry to prove themselves physically equal or superior to other women and men. At the same time, young women from lower socioeconomic classes may find less to lose while weighing warfighting's risk of death and serious injury.

Digital Culture and the Computational Marine

In examining the U.S. Marines' optics of combat in the Iraq War, this book posits a generational difference from previous fighting forces; whether a difference in degree or kind, the 2003 invasion and subsequent occupation are marked by an eye for battle that can be characterized as *synthetic*: a composite system of seeing that coupled high-tech sensors for killing from ever-greater distances with a visual culture of motion pictures and video simulations that eroticized and stylized the action and violence of ground war. Chapter 5's treatment of posthuman warfighting charts the military's increasing utilization of drones and other remote weaponry to the extent that the conflicts in Iraq and Afghanistan may have been the last major wars where the United States relies on a predominantly human ground force. "Think of us as moving from the citizen's army," as historian Tom Engelhardt writes, "to a roboticized and, finally, robot military—to a military that is a foreign legion in the most basic sense."[10] Maybe so, although human perception, coordination, and rapacity in battle may prove extremely difficult to duplicate fully; and the U.S. warrior class, exemplified by the Marines, may prove reluctant to outsource the challenge, play, and gratification of war. It's nevertheless crucial to recognize, as Engelhardt does, that those humans on the American side who fought in the post–September 11 wars were implementing a way of seeing increasingly congruent with the cameras and sensors that would make up the optics for robots and other automated weapons systems. Synthetic views of war suggest a cyborgian convergence between machine and human while denoting a combination of manufactured methods of seeing that bear a highly functional relationship to lethal force projection in war.

To characterize combat optics during the Iraq War as synthetic is not to say that it replaced a type of vision that had recently or ever been truly real, organic, natural, or unmediated. Seeing—even seeing with the "naked eye"— is always mediated by something. At the most basic level, perspective, social context, culture, and language determine the meaning and content of sight. Also, warfighting has long relied on technological extensions of the human eye. Binoculars, aerial photography, radar, and a rifle's telescopic sight belong on the same continuum of enhanced military vision as night vision goggles, the Dragon Eye unmanned aerial vehicle, the LITENING Pod sensor system, and the thermal sight of an Abrams tank. What's more, we know that the Marine Corps, as with the other armed forces, has long depended on an array of moving-picture content to influence the way troops respond to the exigencies of battle. The Hollywood canon of combat movies, as documented in Vietnam War veterans' memoirs, lured young Americans into wartime military service and idealized an aggressive heroic fighting spirit. In a different genre, since World War I the Marines have used in-house-produced training films such as *Bayonet Training* (1938) and *Hand to Hand Combat in Three Parts* (1942) to help troops learn fighting techniques by modeling them on screen. With the larger-than-life demonstrations absorbed into the mind's eye, one's actions on the training field and in battle would more readily follow textbook technique.

And the modeling of fighting through motion pictures was joined in the twentieth century by increasingly lifelike simulations of combat. Before first-person shooter combat video games, the Marines dug mock trenches in Quantico, Virginia, during World War I, developed live-fire courses for World War II, and built mock South Vietnamese villages during the late 1960s.[11] Recognize that such simulation involved an elaborate visual component, as the training courses were meant to have the *look* and feel of the real thing. The *feel* of combat—the physical action and maneuvers—could be imprinted on the marine's person through bodily cognition, as discussed in chapter 3, or muscle memory. In similar regard, sports would serve as a powerful conduit to warfighting. Organized team sports incorporated the culture, color, and pageantry of the military along with its hierarchized institutional structure. At the same time, sport's physical play emphasized control, discipline, and contributing to a team effort, while deeply embedding skills or techniques of bodily performance, some of which could roughly crossover to military service and combat. As women enter the infantry and take on combat roles, sports' influence in shaping young Americans for military service will continue to be demonstrated.

Despite these continuities between the twentieth century and the post–September 11 era, the Marines' optics of combat during the Iraq War can still be distinguished from that during earlier wars. The rapid development of digital technology in the twenty years after the Gulf War (1991) produced several new systems for targeting, communication, and surveillance, at once render-

Synthetic Visions of War   157

Figure 6.1. Major General Ray L. Smith (retired) examines video feed from the UAV Dragon Eye with the First Marine Division during Operation Iraqi Freedom, March 2003. In addition to greatly augmenting American air power, high-technology optics contributed to superior situational awareness of U.S. ground forces. (Bing West)

ing the enemy more vulnerable to lethal attack while further separating and *mediating* the marines from their human targets on the battlefield or city street. I first came to this point somewhat impressionistically, while reading Bing West's valuable book, *The March Up: Taking Baghdad with the United States Marines* (2003), a vivid war story just as concerned with the optics employed by the I Marine Expeditionary Force as with its weaponry or firepower. West presents a compendium of the infantry's digital culture during Operation Iraqi Freedom: third-generation night vision goggles, an iridium satellite cell phone with an encrypted channel, the thermal sight of an Abrams tank, the Blue Force Tracker of friendly units, the "superior" optics of Marine Light Armored Vehicles, the FalconView PC software program that presented three-dimensional digital imagery of a target area, and the Dragon Eye unmanned aerial vehicle (fig. 6.1). *The March Up* chronicles the grunt's perspective during the first days of fighting— the "jiggling sight picture" of a marine aiming his rifle while running, for instance, and a Marine convoy driving by a truck of dead Iraqis: "the Marines stealing sideways glances, their eyes darting away to check buildings for danger, then sliding back for a second look at fresh death." West himself is an ardent picture taker: "I turned on my digital camera, and the battlefield cooperated with continuous action, kilometer after kilometer, growing in intensity."[12] *The March Up* is a primary source for the phenomenon of watching war.

In working from West's book, I've identified and examined the institutions most responsible for generating the visual dynamic so elemental to the way the Marines fought the Iraq War. The United States Marine Corps is an ideal focus for military cultural history concerned with the equation between sight and violence. More times than not, the Marines have been placed at the leading edge of American force projection. Their role as shock troops involves a long-standing claim of exceptionalism within the U.S. armed forces, a self-reflexive institutional identity that, in turn, depends on and emits a certain perspective, an aggressive outlook, both ocular and attitudinal. At the same time, I've extended my research and analysis beyond the Marine Corps. As a historian of modern U.S. culture, I've drawn close connections between the Marines' optics of combat and the early twenty-first-century entertainment industry, the national and global mediascape, the internet, high-technology sensor and weapons systems, and the military-defense industry complex overall.

The Iraq War began on the heels of significant reorientation of the U.S. military via digital culture in the 1990s. Network-centric warfare sought to utilize information age technology in delivering real-time images and other data from an area of operation to field commanders of dispersed combat units. The Rumsfeld Doctrine that drove planning for the 2003 Iraq attack relied on just such advanced information and communications systems, adding a highly mobile and downsized ground force matched with overwhelming air power. Another development in the 1990s that increased synthetic views of war was triggered by the 1993 battle of Mogadishu. With Mark Bowden's book *Black Hawk Down* (1999) and Ridley Scott's 2001 film adaptation gaining unprecedented register in the Pentagon, the challenge of conventional ground forces fighting in heavily populated urban centers predominated military planning in the decade before the Iraq War, including the creation of simulations and sensors that pictured fighting in city streets. The Marines' large-scale training program, Operation Urban Warrior, included a live six-thousand-troop assault landing in Oakland, California, in 1999, just as the Corps started taking bids for a first-person shooter video game devoted to four-man fire team tactics in urban combat.

While in no way new to the turn of the twenty-first century, synthetic views of war took on a layered quality in the few years preceding Operation Iraqi Freedom. We can think of them making up a set of stylized *lenses* for optimal military effectiveness in geographically, topographically, and demographically distinct environments. They can be thought of as visual segues into the radically distinct experience of ground war. Along with the general proposition that combat optics had evolved to a new state of proliferation and technological sophistication by the time of the Iraq War, we can point to key features of visual culture during the Iraq and Afghanistan conflicts that pose significant challenges to peace and war today and in the years to come:

- *Warrior Subjectivity.* Self-possession is a cornerstone of the warrior ethos. Violence in combat should not be governed by anger, hatred, "battle madness," or any other emotional or visceral impulse to kill and destroy. Likewise, deadly force should not be issued reflexively or automatically. Long and careful deliberation over pulling the trigger may not be realistic in most combat situations, but acting with intent and purpose and, ideally, some measure of mindful discrimination is emphasized in ethically informed conceptions of killing in wartime. The Marine Corps in the Iraq War divulges another dynamic at work, however. The resounding metaphor of the killing switch, as discussed in chapter 4, points to an automated system of force projection, a preexistent source of violence that runs through marines like an electrical current or, in the master metaphor of "light it up," like a beam of light that simply needs to shine on its human target. Here a marine's vision is a conduit of deadly force. To see is to kill.

This formulation becomes more problematic with attention to the synthetic nature of the marine's view of war. Shaped by digitized models of killing, the marine's eye for battle is becoming automated in a manner not unlike the way sensors and other optics serve high-tech weaponry. This study of the Marines' visual culture presents a glimpse of a new human subject among the men and women who serve in the U.S. armed forces. Compared to the U.S. military's interpellation of a marine or a soldier one hundred years ago—that is, the way the military institution addressed the warfighter as a Darwinian subject, a unit within the human species, further identified and addressable by race, sex, and gender—today we find a somewhat generic, cybernetic subject emergent among those deployed by the U.S. armed forces. The transformation is far from complete. But in the broadest terms, one senses a computational model eclipsing an anthropological conception of marines and soldiers and a convergence in perspectives between the human warfighter and automated weapons systems.

The computational model of the human warfighter incorporates the precepts that consciousness is uploadable and that cognition can be patterned by content-specific moving pictures and other textual material stored in the mind and triggered and targeted by the eye—especially an eye for battle stylized by pornographic-like appeals to pleasure and gratification through killing in combat and further calibrated through the sight, range, and firepower of one's weapon. Such synthetic inputs—layered onto the mind and eye—delimit warrior subjectivity, a self-reflexive perspective on one's place and actions in a new, complex, and extremely challenging environment. Within this warrior model, synthetic views of battle are also directed to the individual's body and optical system that coordinate physical movement through time and space and the actions taken in response to that changing environment.

- *Bodily Cognition.* The action and behavior of combat can be effectively remembered in the human body: it's a proposition the Pentagon and the Marine Corps have invested in for some time. Boot camp and infantry school—and really the whole military enterprise of physical discipline and regimented training—are based on the idea of imprinting group cohesion (think of close-order drill) and aggressiveness (pugil sticks and bayonet training) on the troop's body and person. Military simulation follows the same thinking, now with the added digital-age promise of ever more closely matching the virtual environment to geographically and topographically specific areas of operation.

The period from 1995 to 2005 marks a high point of the U.S. military's use of first-person shooter video games for infantry training and recruitment. Compared to Hollywood war movies or YouTube war porn videos, combat video games involve rudimentary decision making, the interactivity of play, and a greater sense of spatial movement in how the player's avatar changes positions in virtual space according to joystick and push button controls. While still providing military recruitment with an invaluable inroad to youth culture, video gaming would be succeeded in military training during the Iraq War by more immersive simulation systems, some large enough to house several fire teams. The University of Southern California's 2004 Flatworld became the prototype for advanced systems that set one's physical person in three-dimensional material space and whose virtual environment is filled with digitized human figures guided by artificial intelligence to pose, among other challenges, shoot–

don't shoot scenarios for troops before deployment to Iraq. The Marines' Infantry Immersion Trainer didn't come close to matching the Holodeck effect of Flatworld, but, relying on virtual reality headsets and goggles, the system still freed troops from inert consoles and video screens to allow them to absorb fighting tactics through their own bodily movement. Coordination of the eye and body with environment would produce a sense of "being there," as the thinking went at the turn of the century, "virtual presence" that bestowed a feeling of familiarity to boots-on-the-ground Iraq deployment and consequently helped to pattern killing and assure combat effectiveness.

Today's simulation training technology continues to focus on combining extensive visual representations of likely combat environments and locations with full-body immersion and movement through physical space. In partnering with Northrup Grumman, for instance, the U.S. Army announced in 2014 the Virtual Immersive Portable Environment or VIPE Holodeck, a 360-degree system for training "warfighters in a virtual environment that closely replicates the actual environment they will experience in operational missions."[13] Also in 2014, the Army's National Simulation Center introduced the Future Holistic Training Environment Live Synthetic, an extremely ambitious project that turns on a post–Google Earth digital mapping of the globe so that unit commanders can pull up a discrete location—including the built environment of a particular city—and remotely run his or her troops through that topography. Dubbed Terrain 2025, these data would make up the "digital dirt where," as Lieutenant Colonel Jason Caldwell of the National Simulation Center's Futures Division describes the idea, commanders "can rapidly construct a scenario and get a feel for the terrain that they're going to operate on before they deploy." The Terrain 2025 digital mapping would be only one component within this truly synthetic simulation system. "Those enemy jets are being piloted a thousand miles away by fellow brigade combat team Soldiers, some in aircraft simulators and others on computing gaming stations," is how a recent Army National Simulation Center press release envisions it:

> The Soldiers see the visual recreations of those jets in real-time through special glasses that allow them to see the real world around them, while simultaneously viewing the simulations. Data from the simulations stream in to the Soldiers' glasses from satellites and ground relay stations. In turn, the pilots in simulators and those using gaming stations see what Soldiers are doing in the live environment by satellite and unmanned aircraft video feeds and sensors on the Soldiers

that transmit precise locations and activities. Sounds of the battle are generated through special earpieces that harmonize with the visuals and the smells are pumped in through special odor machines.[14]

Note that amid this flurry of dispersed feeds, relay stations, and sensors, the Army's Future Holistic Training Environment Live Synthetic would still center on the embodied human soldier negotiating a "live environment," albeit one enhanced with topographical and social simulacra carefully selected to represent the sight, sound, feel, and even smell of an otherwise strange place and foreign people upon whom he or she will be expected to project military force. In considering the long arc of military change toward remote and automated weapons, though, the Army's ramped up investment in high-tech simulation to guide the still-human infantry for twenty-first-century combat is itself of positive historical note. It can be taken as a modest investment in humanity and human warfighting. It suggests that the action of effective combat will remain based in somatic knowledge: cognitive embodiment obtained through physical training and muscle memory and deployed through the human eye that heretofore has afforded a social and interpersonal perspective containing compassion and restraint as well as pugnacity and bloodlust.

- *Barack Obama with Remote and Automated Weaponry.* The use of weaponized drones is a profound question of our age, although to call it a question might obscure the already difficult-to-quantify surge in their missions, strikes, and kills since the end of George W. Bush's presidency in 2008 and the beginning of Barack Obama's two terms in 2009. Widespread use of weaponized drones is a fact of U.S. military force today, one that exceeds standard definitions of war in that it is risk free to those who control the drones. It's also a fact of Barack Obama's administration and person. Intellectual biographies of Obama the man will need to take up his transformation from a liberal constitutional law professor who opposed torture and the Iraq invasion to a wartime commander in chief who designed and heads a "nominations process" for remotely killing American and foreign "terrorists" before making any effort to arrest or detain the target.

That Obama's secret program runs free from legal oversight or any other constraint except his personal prerogative goes beyond the ingenuity of his lawyers and even the uber-power of the imperial presidency. Obama sits at the apex of the technological imperative in U.S. military history at a time when the American people have never been more indifferent to whether lethal force is used in their name—just as long as the operation doesn't mean more casualties among their own. Add to that the seemingly permanent casus belli from September 11, 2001, to kill terrorists along with a new remote weapon that increases air power's capacity from destroying whole cities with high-altitude bombing to pinpointing a single human target by airborne camera and ending that person's life with the push of a button, and a reasonable case can be made that Obama enjoys a godlike power unsurpassed in history. Historians will want to consider what happens to the humanitarian sensibility of a progressive politician who can kill whom he can see, when he sees (with some time and effort) almost everyone, and when he may actually gain political support (or less criticism) by erring on the side of shooting. The moral calculation involved in Obama's hit-list program is staggering and finally untenable. One at least hopes that accounts of his anguish over choosing targets are true.

At the same time, when considering the great drift of U.S. military force toward posthuman warfighting, Obama's presidential power and historical input need considerable qualification. The cyborgian convergence of human soldier and fighting machine continues apace. The July–August 2014 *Popular Mechanics* cover story on the new American infantry reports that human sight will be further enhanced by night vision contact lenses and one's leg strength will be boosted by an electronically controlled Tendon-Assisted Rigid Exoskeleton placed between the soldier's heel and ankle. The Army's soldier and the Marines' grunt will also be carrying a smart-scope rifle replete with a laser ranger finder transmitter and receiver and a networked tracking scope.

Moreover, historians will undoubtedly have President Obama following more than leading the military's application of robotics technology in researching and developing automated weapons systems for early to mid-twenty-first-century combat. As with the advanced simulation training systems, the "robot revolution," as stated in the *Popular Mechanics* article, is sold as a cost-saving measure in a time of looming belt tightening: "While automated systems are expensive to develop, they're cheaper to support than humans—they don't need training, healthcare, or condolence letters." This budgetary argument for combat robots is often joined by a quite different consideration: expanding the U.S. military's already glaring asymmetry of power has achieved a type of assumptive force. "Future unmanned systems must be more autonomous," as Air Force chief of staff General Mark Welsh III asserted in 2014, "placing less demand on communications infrastructure and shortening decision cycles."[15]

Further quickening the delivery of lethal force is imperative, as it supports nothing less here than relinquishing human dominion over wartime violence.

Another good reason for combat robots is the withering tolerance the American public has for sacrificing its own people in war. This factor has certainly influenced the content of President Obama's efforts since August 2014 to weaken ISIS in Iraq and Syria with military strikes. Having been elected in 2008 on the promise to end the Iraq War, Obama wants to avoid another land war there as a matter of historical record and practicality. Boots on the ground would bring U.S. casualties, and so Obama has limited the operation to air power even as it proves unable to halt ISIS advances. What percentage of the U.S. sorties have been carried out by weaponized drones is difficult to determine amid the tamped down press coverage in late summer and fall 2014. But Predator and Reaper drones armed with Hellfire missiles have obviously been used widely in what Obama would like to consider post–Iraq War operations. International human rights groups claim civilian deaths from U.S. fire from the air, including eight Syrians in late September from an attack on Raqqa.[16] As of mid-October 2014, the United States has lost one military serviceperson in the operations, a marine who died in an aircraft accident over the Persian Gulf. In early October President Obama escalated attacks against ISIS by ordering low-and-slow-flying Apache attack helicopters into battle against enemy ground troops. "Boots in the air" is how some military experts label this attempt to bring greater force to bear against ISIS while keeping to Obama's air-only pledge.[17] More moral calculus confronts the president as he tries to balance the risk of American casualties against the damage that ISIS is doing in Iraq and Syria. He must be tempted to "send in the Marines!"

- *The YouTube Effect.* Founded in 2005 and bought by Google in 2006, YouTube gained its prominence in American culture through war porn from Iraq and Afghanistan along with goofy cat clips and music videos. Motion picture is the common currency of this video-sharing website: an obvious and redundant statement, but the point is worth underlining in summing up early twenty-first-century visual culture and the human eye's direction and recording of the experience of war and combat. The continuing cultural and cognitive primacy of viewing motion picture is sometimes overlooked in discussions of the digital age. Yet, more than ever, we live in a world composed of moving pictures. Order-of-magnitude expansion of broadband and the internet has sped video's global diffusion through social media platforms like YouTube, Facebook, Live-

Leak, and reddit. Feeding those repositories, though, has been the proliferation of inexpensive and nearly effortless means for recording moving images in the first place. With camcorders dropping below $150 and every smart phone a movie camera—and with minute wearable cameras now taking hold—the twenty-first century has ushered in a digital democratization of perspective and, less positively, the twin impulse to shoot every experience and to feel that it is not truly an experience until it is so shot. The existential conflation today between *being* and *possessing footage* of being is laid out nicely in journalist Nick Paumgarten's article on the GoPro camera and the small, often-head-mounted digital camera's effect on mountain biking and other adventure sports. As getting the shot of a stunt or a bike run supersedes the action itself, "the adventure athlete reveals himself to be something else: a filmmaker," Paumgarten writes, "a vessel for the creation of content."[18]

Exactly what this finding means for today's warrior culture is just beyond my study's coverage. We have seen in chapter 2, though, that U.S. marines during the first years of the Iraq War were already adept takers of motion pictures, some with rigged-up helmet cams, others with hand-held camcorders and Flip cameras. The social history of U.S. ground forces in Iraq waits to be written. Use of personal computers and basic video-editing software will need to be included in studies of infantry life in camp. As grunts and soldiers produced and shared videos from their combat footage, we know they shaped the early content, growth, and culture of YouTube. Whether marines aggrandized their actions in the "field" for better video content is another matter. I found no evidence in early war pornography from Iraq of marines, unlike Paumgarten's mountain bikers, being motivated by "virality." At the same time, and in a much broader sense, warfighting has long incorporated the promise of recording one's heroic actions in memory, both vernacular and official, and representing them after the fact in story, image, commendation, and commemoration. Stephen Crane's novel *Red Badge of Courage* makes plain the performative essence of martial heroism: "a blue determined figure standing before a crimson and steel assault, getting calmly killed on a high place before the eyes of all." The soldier imagined "his dead body" exciting a "magnificent pathos" on the battlefield, and "these thoughts uplifted him." In a more quotidian vein, marines in Iraq realized that fighting equaled fighting footage. During the first decade of the twenty-first century, motion picture from Iraq went into what I've described as a tightening cultural loop between the conduct of battle and its visual record; those fighting in Iraq may have been influenced by earlier combat films, while videos from their own action composed a kind of war pornography consumed by would be marines on the internet.

The virality of motion picture also raises the very timely point that the United States is unable to completely control what motion picture its enemies upload to the internet and thereby enter into contemporary visual record. Comprehensive discussion of the global mediascape's openness to (from the perspective of U.S. security) counter content would need to take up the visual fact of the September 11 attacks themselves and examine the common contention that, per the logic of terrorism, they were undertaken more for the eyes of the world than to weaken American defenses by lessening its population. Al Jazeera's coverage of the Global War on Terror from a myriad of viewpoints would warrant treatment; as would the April 2003 U.S. missile strike on the network's Iraq bureau that killed one reporter. Primary focus would be placed on ISIS's sophisticated use of social media and motion picture production. The videotaped beheadings of U.S. journalists James Foley and Steven Sotloff in late summer 2014 would only be the most apparent examples. In putative response to U.S. airstrikes against Islamic State forces in Iraq and Syria, the beheading videos were meant to provoke the United States and draw it further into war. Islamic State war pornography has also been effective recently in recruiting new fighters from the Persian Gulf region and the West.

The internet can be understood to be its own front in the war against radical militant Islam. Less a medium or coordinator of force, as in network-centric warfare, digital culture and global communication together make up a contested site, a dynamic yet finite context for transferring message, meaning, and cognition. Under this view, though, the finitude isn't really the digital platform itself, whose expansion through technological development seems boundless, but the time, attention span, and cognition of the world's population using the Internet and related digital media. In terms of visual culture and motion picture, we may be headed toward everything and everyone being filmed and shared from nearly infinite digitally recorded perspectives. But no one would be able to come close to taking it all in. Practical decisions must be made as to what to watch, while different interests and powers try to draw people to its content and limit their access to others. With this in mind, the internet and visual culture as battlefront melds with an older notion of a marketplace of images and ideas. Indeed, this whole book on the U.S. Marines in the Iraq War and the optics of combat can be understood as an intervention in the political economy of vision, one necessarily global in its input and reach, and one whose currency helps channel and project military force and organized violence. Under this view, then, an institution like the United States Marine Corps seeks to fill up and control its warfighters' visual environment. Synthetic views of war are layered onto a marine's sensibility by what some theorists would call an optical *regime*, a field of power that is intimately tied to and calibrated with the Corps' "killing switch" and its effectiveness as an assault force.

Subjectivity Lives and Dies

Talk of all-encompassing synthetic views of war brings us full circle to the tension between automated killing and warrior subjectivity: defined simply as the sense of individual self, subjectivity is the source of ethical responsibility in war—maintaining dominion over one's mind, as discussed in chapter 4—as well as the kernel of personal autonomy that accounts for biographical continuity ("I'm still me") through war's ravages and military training's "brainwashing." Subjectivity lives. It's evident throughout the preceding chapters. Programmability of a human mindset functional to killing in combat is a military work in progress, one that nears its goal in shades only, even as the turn of the century brought an extremely deadly and highly crafted (pornographically and technologically) eye for battle to the Iraq War. U.S. marines and soldiers deployed to Iraq and Afghanistan started to record their experiences in images and words the day they got there; and for those who survived their tours, that effort continues in interviews, letters, memoirs, blogs, speeches, discussion panels, and storytelling. It's an increasingly rich repository of subjective thought, feeling, and sense impression among young men and women who may have made up the last major all-human ground force in U.S. history. In having the privilege of being one of the first professional historians to work in this material, it's crucial to make clear that to say "warrior subjectivity lives" is to recognize, first, the crisis in mental health among today's veterans and, second, that, for many, subjectivity becomes the repository of the guilt and shame over killing other human beings. "Self-possession" doesn't roll off too many recent combat veterans' tongues. Self-respect is often a casualty as well. For far too many veterans, self-understanding elides with self-hatred and self-destruction in many forms.

Examples abound. Sergeant Colin Keefe, the resolute machine gunner we met in chapter 1, certainly suffered for years after his fighting in Operation Iraqi Freedom. Corporal Tom Lowry, the broken man discussed in chapter 4,

Figure 6.2. James Blake Miller, First Battalion, Eighth Regiment, during the second battle of Fallujah, November 2004. This photograph, taken by *Los Angeles Times* photographer Luis Sinco, made Miller into a type of celebrity. Medically discharged from the Marines in November 2005, Miller would later speak of not knowing why he survived a war that took so many other lives. (*Los Angeles Times*. All rights reserved)

died in a spring 2014 automobile accident late at night on an empty road in North Carolina. Or consider James Blake Miller, whose young and weary visage photographed during combat in Fallujah in 2004 became one of the most iconic images from the Iraq War (fig. 6.2). This is one last image in my study of images, moving pictures, and the optics of war. With Miller's eyes staring one thousand yards, his face streaked with mud and blood, a cigarette hanging from his lips, the Marlboro Man portrait celebrated, for many in late 2004, only a year and a half into the war, the hardened ferocity of the American warrior and, indeed, of the U.S. marine.

The postscript to Miller's image and service tells a different story or, rather, a latter part of the same story. In an accomplished 2007 documentary short by Chad A. Stevens, we find Miller back home in rural Kentucky, medically discharged from the Marines and, in his words, "confused as to why I'm still here and why other people aren't." His continued existence is quite uncertain.

Without any "promise of tomorrow," Miller says, "any day above ground is a good one." Constantly thinking about Iraq, Miller "didn't want anything to do with anybody." And with few meaningful social relations, "I had more or less planned out my death," Miller continues: "I wanted to get rid of the pain. . . . I was actually kind of happy for a moment thinking that I wasn't going to feel anything soon."[19] Miller persevered, but his thoughts elucidate the suicide epidemic among U.S. military personnel and veterans.

The mother of an U.S. Army veteran of the Iraq War who killed himself after discharge explains what happened: "He thought of himself as a murderer, and a bad person, because he still had the urge to hurt people. The United States Army turned my son into a killer. They trained him to kill to protect others. They forgot to un-train him, to take that urge to kill away from him." Her son had not been physically wounded in Iraq. But he came back, nonetheless, in mortal danger. His kind of wound, she says, "kills you from the inside out." As military suicide rates rose precipitously during the Iraq and Afghanistan wars—peaking in 2012 with thirty-three suicides per month among U.S. military personnel and, among veterans, twenty-two per day in February 2013—psychologists and others working with veterans returning from war have come to see combat trauma as less a disease or disorder and more as an injury.[20] A pivotal figure here is clinical psychiatrist Jonathan Shay, who coined the term *moral injury* to denote the violation in combat of an ethical belief or moral code (i.e., "Thou shalt not kill") resulting in severe injury to the soul or psyche. "It is," as classicist scholar Robert Emmet Meagher has recently written, "what used to be called sin."[21] Overwhelmed by guilt and shame, the eating away of the combat veteran's soul and self presents another really good argument for weaponized drones and robots or for less war.

James Blake Miller's "Marlboro Marine" portrait is historically significant for what it tells us about his vision during combat. Consider the photograph again. He may be staring off into space during the break, but he's far from deranged, catatonic, or otherwise incapacitated. If we move our focus from his visage to his vision—from what he looks like to what his point of view is—we can well imagine, extrapolating from our knowledge of Fallujah in November 2004, what will soon be his poise, focus, and deadliness in fighting. Miller's own words from the documentary short offer an extraordinary window into his eye for battle, the grunt's homicidal resolve, and the doomed condition of the marine or soldier who does his or her job. "My nightmares revolve around a lot of faces," is how Miller begins to detail the "re-occurring dream" at the center of his Iraq War memories. It's "me actually looking down the rifle and knowing that I'm going to pull the trigger on that person," Miller continues: "It's an insane connection that you make with the person at that point. To see somebody in your sights, and to pull that trigger. It's almost like you're there with them, seeing their life flash before your eyes, as well as taking it."[22] While such clearly stated comprehension is rare in the U.S. Iraq War record, setting one's rifle sight on "somebody," perceiving the human "connection" with that "person," and pulling the trigger is arguably the most characteristic sequence of actions in modern warfare. Taking the shot regardless of that human connection is at once a requisite for war to continue in its current form and an unnatural behavior, according to the military's own psychology, an "insane" moment, in Miller's word, as it ends the life of another human being and, for many, goes on to haunt the one who walks away.[23]

It would be a mistake, though, to think of every war veteran as haunted or doomed. Veteran and author Eric Fair insists on the social imperative not to pathologize everyone who has served in war. Subjectivity lives. Just as there's no one military mindset functional to killing, there's no single psychological or moral fate awaiting one who has been in battle. It's more complicated than that. And for my money, reading and writing are still the most promising behaviors for coming to robust understanding of war. As immersed as we are in motion picture, this is not, as some claim, a postliterate time or culture—which is a long way to say that I want to end this book by privileging the literary eye for battle and its capacity for gaining some critical descriptive distance on the experience of war. The imagination that goes into both reading and writing about war sets literature apart from other cultural forms. To be sure, a great many pages are formulaic, pornographic, and generally wanting. The immense popular literary culture surrounding the U.S. military and its history isn't exactly calibrated for coming to new hard truths about war. Too often just the opposite. By the same token, literature's dynamism is too often overlooked. The written word has the power to prompt in the reader's imagination and record in the mind's eye vivid landscapes of war, battle scenes that are rendered and absorbed not statically but in motion picture. "As the regiment swung from its position out into a cleared space the woods and thickets before it awakened," Stephen Crane wrote in *Red Badge of Courage*: "Yellow flames leaped toward it from many directions. The forest made a tremendous objection."[24] The written word contributes fundamentally to the visual culture of war.

Another dynamic attribute of war reading and writing is the ironic mode. Seeing through the gung ho excesses of military indoctrination and culture enjoys a long and rich tradition among American soldiers and marines. Colin Keefe speaks about the clear difference between the few "brainwashed ones" in boot camp, infantry training, and deployment who "live and breathe for the Marines," and "my kind," who, for example, would never utter "Semper Fi" or "Hoo rah" on their own accord.[25] If subjectivity lives, it does so by sizing up a master institution like the Marine Corps and countering the full weight of its purpose and practices by letting them wear only lightly on your person. Irony in war literature centers on a similar sensibility of knowing better. In *The Great War and Modern Memory* (1975), literary critic Paul Fussell defined battle-borne irony as the "collision between innocence and awareness"—a collision resulting in deflated darkness and disillusionment (or worse) for most who experience it, but also an energy source for examining in poetry and prose the duplicity of war's promises and ravages.[26]

The play of mind between subject positions and perspectives is a strong suit of war writing. Within the Marine literary canon, there's no better example than Anthony Swofford's Gulf War memoir *Jarhead* (2003), a work we've already encountered in chapter 2's discussion of Hollywood combat movies' function as pornography for the marine preparing for battle. Through first-person narration, Swofford sets the reader in the middle of the drunken "Vietnam War Film Fest" on base in August 1990, when his unit had just been told of "imminent war" against Iraq. He then adds a second perspective on the scene by stepping back and explaining its import and meaning from his sober-minded view well after he had left the Marines. "Filmic images of death and carnage are pornography for the military man," Swofford writes: "With film you are stroking his cock, tickling his balls with the pink feather of history, getting him ready for his real First Fuck." Fast forward to the Gulf War's last hours and Swofford the Marine sniper set in a hide close to an Iraqi-controlled airstrip. Primed on Hollywood war porn, Swofford fantasizes pulling his rifle's trigger, "hopping from head to head with my crosshairs, yelling, *Bang, bang, you're a dead fucking Iraqi*." Somewhat similar to James Blake Miller's nightmares over the "insane connection" between shooter and target, Swofford the marine pontificates sardonically:

> The enemy are caught in an unfortunate catch-22, in that I care for them as men and fellow unfortunates as long as they are not within riflesight or they're busy

keeping dead. But as soon as I see them living, I want to perform some of the most despicable acts I've learned over the prior few years, such as trigger killing them from one thousand yards, or gouging their hearts with my sharp bayonet.

Jarhead's critical power comes in showing the pathological desire to kill to be the flip side of war pornography. Swofford aches to shoot; it's both want and need. And as he continues to work himself up by imagining what his rounds would do to the human in his sights—Swofford pictures the "pink mist" resulting from a perfect head shot—we understand all the more clearly how the pathology to kill remains instilled in the eye.[27]

While the literary canon from the Iraq and Afghanistan wars is still taking shape, Marine Corps veteran Phil Klay's short-story collection *Redeployment* (2014) will undoubtedly be placed close to the top. National Book Award winner, the collection's twelve stories render a complex of subjects and perspectives on early twenty-first-century wartime America and the yawning gap in understanding and concern between those fighting in Iraq and Afghanistan and those who are not. Army veteran Kevin Powers's novel *The Yellow Birds* (2012) has also received considerable critical acclaim. Powers's work is, first and foremost, a literary effort. Told from the perspective of an eighteen-year-old private trying to stay alive during his first infantry tour in Iraq, the book presents well-crafted thought pictures and sense impressions from the front. "Through the daylight hours we took turns on watch, sleeping for two hours and nodding off behind our rifles for one," Powers writes: "We saw no enemy. We made up none out of the corners of our eyes. We were too tired even for that. We saw only the city, appearing as a blurred patchwork of shapes sketched in white and tan beneath a ribbon of blue sky." Very well written, Powers's prose also bears a requisite knowingness or irony in the private's sensibility. As his unit completes its preparation for deployment to Iraq, the soldiers hold a celebratory dinner with family in the base gymnasium. A satirical note can be heard in Powers's Hemingway: "And there were hamburgers and French fries and we were glad."[28] It's a thoughtful book. But *Yellow Birds* does not have the psychological and emotional immediacy found in the most memorable and effective war writing: from Crane to Sledge to Swofford and now to Klay, especially as seen in his title story "Redeployment."

This story of a marine coming home from Iraq turns on a multiplicity of perspectives, all of them emerging through the spasmodic movement of his own sight and mind's eye. Sitting with his platoon on a commercial airliner—

"everybody's hollow-eyed and their cammies are beat to shit"—Sergeant Price "look[s] up at a little nozzle shooting air-conditioning, thinking, What the fuck?" Price tries to get a handle on things. "You close your eyes, and think. The problem is, your thoughts don't come out in any kind of straight order." He continues:

> You don't think, Oh, I did A, then B, then C, then D. You try to think about home, then you're in the torture house. You see the body parts in the locker and the retarded guy in the cage. He squawked like a chicken. His head was shrunk down to a coconut. . . . You see the things you saw the times you nearly died. The broken television and the hajji corpse.[29]

Amid the jumble of flashbacks crashing through the simple narrative of settling back into living with his wife, "Redeployment" is framed by substories about shooting dogs, beginning with feral dogs in Fallujah and ending with his beloved but suffering dog at home, Vicar. From the mindless "open season" on Iraqi canines to putting down the old and dying Vicar, one finds an arc toward intent and purpose, or at least of Price gaining an initial hold of himself, even if it is via the expertise of shooting and killing imprinted on his person by the Marines. You have to start somewhere. "Sight alignment, trigger control, breath control. Focus on the iron sights, not the target. The target should be blurry." It's the one still and focused moment during Price's return. While aiming at Vicar's "gray blur," Price pulls the trigger three times, "each shot coming quick after the last so you can't even try to recover, which is when it hurts." Price's memory of fighting and killing in Fallujah isn't nearly as optically succinct. There's no focus or coherent sequence of images. "And it's not easy to kill people either," Price blurts out to himself while still flying home. But then he immediately sees in memory his company's executive officer effortlessly slitting the throat of a dying insurgent with his KA-BAR, exclaiming "It's good to kill a man with a knife." Klay devotes "Redeployment" to the veteran's effort to calibrate the appearances of a slowly emerging reality back home with the imagistic bursts from war. It's a more optimistic story than some acknowledge. The colors in Price's eye and feelings can be trusted, sometimes. Upon finally reuniting with his wife at Camp Lejeune, Price embraces her and notices how well her body "fit into mine." Such a "new feeling," Klay writes, "made everything else black and white fading before color."[30]

Subjectivity would come in fits and starts. The story's centerpiece is a shopping trip Price takes with his wife to nearby Wilmington, North Carolina, where walking down the city's street and ducking into stores to try on clothes are suddenly overlaid by visual jolts of patrolling Fallujah: "In a city there's a million places they can kill you from. It freaks you out at first. But you go through like you were trained, and it works." The problem is that "you don't have a squad" in Wilmington, "you don't have a battle buddy, you don't even have a weapon." Price tells himself he's safe. Instead he "goes to orange": intense alertness experienced through the eye, a colored optics that only comes "in a firefight, or the first time an IED goes off that you missed." Some marines lose it in combat, Price says in explaining his condition. They "go straight to red. They stay like that for a while and then they crash, go down past white, down to whatever is lower than 'I don't fucking care if I die.'" Otherwise, though, orange is the norm for boots on the ground in Iraq: "Most everybody else stays orange, all the time." But orange doesn't work for Price in American Eagle Outfitters—how can people casually walk by those windows? The desperate incongruity between Price's high alert and peaceful Wilmington provides a climax to Klay's story. The "amped-up" Price and his wife beat a hasty retreat. She won't let him drive: "I would have gone a hundred miles per hour."[31] Once back from Wilmington, they together take in Vicar's misery and decide to end his life. It's a good ending to a superb story. Orange gives Klay the alarming blare he wants for coloring Price's instability. Visual fields from Iraq barge into his consciousness; far from self-possessed, he's unable to control his own senses.

Klay's orange adds nicely to the historical record of combat optics in the Iraq and Afghanistan wars. Before heading home from Wilmington, still stuck in that city's streets, Price defines what the color meant for him and his platoon in Fallujah: "Here's what orange is. You don't see or hear like you used to. Your brain chemistry changes. You take in every piece of the environment, everything. I could spot a dime in the street twenty yards away. I had antennae out that stretched down the block." Orange amounts to a typology for "amped-up" sight and perception in war. Price doesn't say whether the color changes in combat, although orange contains both the bright of alertness and the red of violence.

Compared to other leading literary views of war, there's a steadiness and certainty in Klay's orange. It's no "battle madness," as found in Stephen Crane's war hero; there's no "strong personal hate" or "cold, homicidal rage" that drove E. B. Sledge at Peleliu and Okinawa; and there's nothing suggesting Anthony Swofford's psycho killer. Klay's orange is more like a sensor system—a synthetic eyeball. It shades into an automated view of war. Still in Wilmington, Price describes the color orange as nothing more or less than intense mental computation. With ground conditions in Iraq constantly changing, "you take on too much information to store so you just forget, free up brain space to take in everything about the next moment that might keep you alive. And then you forget that moment, too, and focus on the next. And the next. And the next. For seven months." (No wonder combat video games are used in cognitive therapy for veterans' post-traumatic stress; the wartime layers of vision that still color reality after Iraq need to be peeled back slowly; first-person shooters modulate that process, pacifying veterans with familiar snippets of virtual war.) As "Redeployment" demonstrates, one adapts gradually, if at all. Although, again, Klay's lead story about the chasm between Iraq and home bears hope and connection. Price starts to see a situation's color for what it is. After shooting his dog, "Vicar was a blur of gray and black."[32] The world is no longer constant orange.

I keep picturing the four young men in uniform from Quantico's Officer Candidates School at the U.S. Marine Corps museum on July 4, 2012. The Belleau Wood film's close-ups of marines getting mowed down by German machine gun fire may have been startling, but the four candidates probably took greater note of the Americans killing Germans while advancing on their line. Killing, not dying, is the essential act of war. Wartime killing is a source of both personal gratification and self-injury—to the body, mind, and soul. In addition to the valuable clinical and spiritual work being done with veterans' post-traumatic stress and moral injury, I wonder if a longer-term historical adjustment could be initiated to make sure that young men and women considering military service really understand its risks and dangers. It's not as if they don't have reason to know any better. The moral "violation" of war is a well-worn truth. It has a long and well-documented history. And yet veterans today utter shock and dismay over its continued occurrence. Take Phil Klay's Sergeant Price and his lament over the seven months in Iraq at orange. If I could speak with Price I'd ask him about it. Not as historian to subject during an interview, but as a fellow citizen over coffee or beer. What in the hell did you think a tour of duty with the U.S. Marines would be like? How in the world could moral injury come as a surprise? It's commonplace to note the intergenerational cycle of forgetting chronic with the horrors, death, and loss of war. It's a lesson, tragically, in continual need of relearning from experience rather than from understanding the past. But that's not good enough. If the United States is to have a warrior caste, then it should be one with eyes wide open to the degradation and destruction that military force necessarily brings to one's self and others. Volunteering for military service is a political act. If warrior subjectivity lives, then it manifests individual choice and responsibility for itself.

Notes

Acknowledgments

1. Michael Sherry, *The Rise of American Air Power: The Creation of Armageddon* (New Haven: Yale University Press, 1987), 1.

INTRODUCTION: Force Projection and the Marine Eye for Battle

1. John Pettegrew, *Brutes in Suits: Male Sensibility in America, 1890–1920* (Baltimore: Johns Hopkins University Press, 2007).

2. Number of U.S. casualties in the Iraq War comes from the U.S. Department of Defense. http://www.reuters.com/article/2013/03/14/us-iraq-war-anniversary-idUSBRE92D0PG2013 0314. http://www.defense.gov/news/casualty.pdf (accessed June 5, 2014); "virtually countless Iraqi dead and wounded" references the U.S. military's stated policy of not doing body counts for the Iraq War and the great uncertainty over numbering Iraqi casualties from the war—a war that, for Iraqis, is very much ongoing, with mounting dead and wounded. The Iraq Body Count project, known as being on the conservative end of the spectrum in estimating Iraqi deaths, charts "the public record of violent deaths following the 2003 invasion of Iraq." Its website, on June 19, 2014, lists "documented civilian deaths from violence" between 126,086 and 140,933; and "total violent deaths including combatants" at 191,000. It further details 33 killed on June 18, 2014, and 2,764 civilians killed in June 2014. https://www.iraqbodycount.org (accessed June 19, 2014).

3. Franks quoted in Michael R. Gordon and General Bernard E. Trainor, *COBRA II: The Inside Story of the Invasion and Occupation of Iraq* (New York: Pantheon Books, 2006), 110. In addition to *COBRA II*, for the invasion I've relied on Thomas E. Ricks's valuable *Fiasco: The American Military Adventure in Iraq* (New York: Penguin Press, 2006); and Ali A. Alawi, *The Occupation of Iraq: Winning the War, Losing the Peace* (New Haven: Yale University Press, 2007).

4. For the autonomy and strength of Marine culture, Thomas E. Ricks, *Making the Corps* (New York: Touchstone Books, 1997) is particularly useful: "But the Marines are distinct even within the separate world of the U.S. military. Theirs is a culture apart. The Air Force has its planes, the Navy its ships, the Army its obsessively written and obeyed 'doctrine' that dictates how to act. Culture—that is, the values and assumptions that shape its members—is all the Marines have. It is what holds them together. They are the smallest of the U.S. military services, and in many ways the most interesting. Theirs is the richest culture: formalistic, insular, elitist, with a deep anchor in their own history and mythology. Much more than the other branches, they place great pride and responsibility at the lowest levels of the organization. The Marines have one officer for every 8.8 enlistees. That is a wider ration than in any other service—the Air Force, at the other end of the spectrum has one officer for every 4.2 enlistees" (19). Essential is Aaron B. O'Connell, *Underdogs: The Making of the Modern Marine Corps* (Cambridge, MA: Harvard University Press, 2012). See also Craig M. Cameron, *American Samurai: Myth, Imagination, and the Conduct of Battle in the First Marine Division, 1941–1951* (New York:

Cambridge University Press, 1994). The authoritative overview of the U.S. Marine Corps is Allan R. Millett, *Semper Fidelis: The History of the United States Marine Corps* (New York: Free Press, 1980).

5. "Trained instinct" often appears in modern Marine discourse, including Bing West, *The Strongest Tribe: War, Politics, and the Endgame in Iraq* (New York: Random House, 2008), 57.

6. S. L. A. Marshal, *Men against Fire: The Problem of Battle Command* (New York: William Morrow, 1947); Dave Grossman, *On Killing: The Psychological Cost of Learning to Kill in War and Society* (New York: Back Bay Books, 1995), 31, 253.

7. For a valuable summary of the criticism leveled against Marshall's study, see Peter S. Kindsvatter, *American Soldiers: Ground Combat in the World Wars, Korea, and Vietnam* (Lawrence: University Press of Kansas, 2003), 220–27.

8. Ricks, *Making the Corps*.

9. O'Connell, *Underdogs*, 5.

10. Chris White, "First to Fight Culture: A Former Marine on the Marine Motto," *Counterpunch*, May 29, 2004.

11. "Recruit Training" at www.Marines.com, http://www.marines.com/becoming-a-marine/recruit-training (accessed June 14, 2014).

12. http://www.pendleton.marines.mil/StaffAgencies/AssistantChiefofStaffG35/TrainingSupportDivision/MOUTFacilities.aspx (accessed May 23, 2013); http://usmilitary.about.com/od/usmcbase/ss/TwentyninePalms.htm (accessed May 23, 2013).

13. Gordon and Trainor, *COBRA II*, 183–84.

14. Bing West, *The March Up: Taking Baghdad with the United States Marines* (New York: Bantam Dell, 2004), 11–12.

15. Lieutenant Colonel Bryan P. McCoy, "Understanding the Battlefield," *Marine Corps Gazette* (November 2005): 42–44.

16. West, *The March Up*, 16.

17. Paul Virilio, *War and Cinema: The Logistics of Perception* (London: Verso Press, 1989).

18. This foundational question of military cultural history is from John Keegan, *The Face of Battle* (New York: Penguin Press, 1976), 16.

19. Ricks, *Making the Corps*, 22.

20. David J. Danelo, *Blood Stripes: The Grunt's View of the War in Iraq* (Mechanicsburg, PA: Stackpole Books, 2003), 9.

21. Marco Martinez, *Hard Corps: From Gangster to Marine Hero* (New York: Random House, 2007), 157, 158.

CHAPTER 1: Shock and Awe and Air Power

1. Author audiotape and transcribed interview with Colin Keefe, December 5 and 21, 2005.

2. Michael S. Sherry, *The Rise of American Air Power: The Creation of Armageddon* (New Haven: Yale University Press, 1987), ix, xi.

3. Franks quoted in Adrian R. Lewis, *The American Culture of War: The History of U.S. Military Forces from World War II to Operation Enduring Freedom* (New York: Routledge, 2005; 2nd ed., 2012).

4. These terms, overlapped in meaning and inconsistently used by military policy makers and analysts, benefit from delineation: *transformation*, synonymous with "the Rumsfeld Doctrine," is most closely associated with Secretary of Defense Donald Rumsfeld (2001–6 under President George W. Bush) and his effort to reorganize the U.S. military around high-technology combat systems, air power, and small highly mobile ground forces; *network-centric warfare* emerged as a term in the mid-1990s and centers on a "system of systems" based on satellite

and digital communications dramatically augmenting situational awareness and mission effectiveness; *revolution in military affairs* is an older term, emerging after World War II and also used by non-U.S. theorists, whose usage is broad and diffuse—going to the adoption of advanced weapons and communications technologies, radical reorganization of armed forces' structures, hierarchies, and command systems, and the conception of military strategy and tactics within a global context; *a new way of war*—stemming from Russell Weigley's seminal work in military history, *The American Way of War: A History of United States Military Strategy and Policy* (Bloomington: Indiana University Press, 1973)—is a term used by journalists and military commentators in referring to the high tech orientation of the U.S. military, on which see, e.g., Max Boot, "The New America Way of War," *Foreign Affairs* (July–August 2003), http://www.foreignaffairs.com/articles/58996/max-boot/the-new-american-way-of-war (accessed October 3, 2013).

5. *Immaculate warfare*, in contrast to the terms in the preceding note, is something of a sarcastic or critical term denoting the exaggerated claims of those promoting network-centric warfare and such. For example, Ralph Peters, "The Myth of Immaculate Warfare," *USA Today*, September 5, 2006, http://usatoday30.usatoday.com/news/opinion/editorials/2006-09-05-warfare-edit_x.htm (accessed October 8, 2013).

6. John Hamre, former Deputy Secretary of State, quoted in Tim Weiner, "Pentagon Envisioning a Costly Internet for War," *New York Times*, November 13, 2004, http://www.nytimes.com/2004/11/13/technology/13warnet.html?_r=0.

7. David S. Alberts, John J. Garstka, and Frederick P. Stein, *Network Centric Warfare: Developing and Leveraging Information Superiority* (Washington, DC: CCRP Publications, 2004), 3.

8. Harlan K. Ullman and James P. Wade, *Shock and Awe: Achieving Rapid Dominance* (New York: NDU Press Book, 1996), http://www.shockandawe.com/ (accessed April 25, 2006).

9. S. S. Coer, "Fifth-Generation War," *Marine Corps Gazette* (January 2009): 65.

10. Lewis, *American Culture of War*, 379.

11. Michael R. Gordon, *Cobra II: The Inside Story of the Invasion and Occupation of Iraq* (New York: Pantheon Books, 2006), 300.

12. Rebecca Grant, "Eyes Wide Open," *Air Force Magazine* 86 (November 2003), http://www.airforcemag.com/MagazineArchive/Pages/2003/November%202003/1103eyes.aspx (accessed January 22, 2014).

13. Rebecca Grant, "JSTARS Wars," *Air Force Magazine* 92 (November 2009), http://www.airforcemag.com/MagazineArchive/Pages/2009/November%202009/1109jstars.aspx (accessed December 14, 2013).

14. John R. Thomas, "Transforming Marine Corps C(4)," *Marine Corps Gazette* (August 2002), http://pqasb.pqarchiver.com/mca-members/doc/221456421.html?FMT=FT&FMTS=ABS:FT:TG:PAGE&type=current&date=Aug+2002&author=Thomas%2C+John+R&pub=Marine+Corps+Gazette&edition=&startpage=16-19&desc=Transforming+Marine+Corps+C%284%29 (accessed January 22, 2014).

15. Ullman and Wade, *Shock and Awe*.

16. Ibid.

17. James Wade used the word "really" in saying that Operation Iraqi Freedom followed shock and awe doctrine in Joel Achenbach, "Military Theory and the Force of Ideas," *Washington Post*, March 23, 2003, http://www.washingtonpost.com/wp-dyn/articles/A12208-2003Mar22.html (accessed February 10, 2014).

18. Ullman and Wade, *Shock and Awe*.

19. Lewis, *American Culture of War*, 441; Gordon, *Cobra II*, 69.

20. Lewis, *American Culture of War*, 436.

21. Ullman and Wade, *Shock and Awe*.

22. Gordon, *Cobra II*, 210.

23. "DoD News Briefing—Secretary Rumsfeld and General Myers," March 21, 2003, http://www.defense.gov/transcripts/transcript.aspx?transcriptid=2074 (accessed February 23, 2014).

24. Iraq Body Count, https://www.iraqbodycount.org, includes "A Dossier of Civilian Casualties in Iraq, 2003–2005" (accessed December 10, 2013); Lea Roberts, "Mortality before and after the 2003 Invasion of Iraq: Cluster Sample Survey," *Lancet* (October 2004): 1857–64, http://www.jhsph.edu/research/centers-and-institutes/center-for-refugee-and-disaster-response/publications_tools/iraq/sdarticle.pdf. For summaries and discussion of Iraqi casualties, see Costs of War, http://www.costsofwar.org/article/civilians-killed-and-wounded (accessed January 10, 2014).

25. Robert Pape quoted in John T. Correll, "What Happened to Shock and Awe?," *Air Force Magazine* 86 (November 2003), http://www.airforcemag.com/MagazineArchive/Pages/2003/November%202003/1103shock.aspx (accessed January 20, 2014).

26. Franks quoted in Gordon, *Cobra II*, 211; Lewis, *American Culture of War*, 441–42.

27. Michael D. Visconage, "Into Iraq, 2003: Marine Aviation in the Opening Gambit," *Leatherneck* (March 2013): 34–39; Amos quoted in ibid., 34.

28. Interview of Lt. Colonel Willard A. Buhl, November 4, 2003, Iraq War Oral Histories, United States Marine Corps History Division, Quantico, Virginia.

29. Visconage, "Marine Aviation in the Opening Gambit"; Amos quoted in ibid., 37.

30. Interview of Major Scott Wedemeyer, May 5, 2003, Iraq War Oral Histories. Lieutenant Colonel Kevin Vest also made the point: "It was obviously a tremendous advantage to be able to fly over the country of Iraq basically in the profiles that we were going to use in combat prior to the war ever starting. . . . it knocked off a lot of nervousness amongst the pilots." Interview, April 15, 2003, Iraq War Oral Histories.

31. Wedemeyer, Iraq War Oral Histories. Captain Joshua M. Pieczonka, a weapons systems operator on a F/A-18 Hornet, spoke of how he and others would "utilize our sensors and our eyeballs, most important thing." Interview, May 6, 2003, Iraq War Oral Histories.

32. Pieczonka, Iraq War Oral Histories; Vest speaks of being called off strikes, Iraq War Oral Histories.

33. Lt. Colonel Clyde Woltman, April 15, 2003, Iraq War Oral Histories.

34. "Marines to Get More LITENING AT Targeting Pods for F/A-18s," *Space Daily*, June 14, 2005, http://www.spacedaily.com/news/gps-05zzzb.html (accessed February 25, 2014).

35. Interview with Captain Jennifer M. Dolan, May 13, 2003, Iraq War Oral Histories.

36. Ibid.

37. Chris McPhillips and Michael A. Boorstein, "Digital Targeting in the MAGTF," *Marine Corps Gazette* (May 2005), https://www.mca-marines.org/gazette/2005/05/digital-targeting-magtf; Benedict C. Buerke, "Essential Tools for Effective Urban CAS," *Marine Corps Gazette* (May 2008), https://www.mca-marines.org/gazette/2008/05/essential-tools-effective-urban-cas.

38. Vest, Iraq War Oral Histories.

39. Bing West, *The March Up: Taking Baghdad with the United States Marines* (New York: Bantam Dell, 2003), 67.

40. Dahr Jamail, "Living under the Bombs," in "Tomgram: Dahr Mamail on Life under the Bombs in Iraq," February 2, 2005, http://www.tomdispatch.com/blog/2166/.

41. Dexter Filkins, *The Forever War* (New York: Vintage Books, 2008), 122.

42. Author interview with Colin Keefe, January 2014, from which these quotes are taken.

43. Keefe 2005 interview; I describe guilt as longer lasting because the point dominates his 2014 interview.

44. Keefe 2005 interview.

45. West, *March Up*, 144. Interview with Sergeant Joshua Laverty, December 12, 2003, Iraq War Oral Histories: "Iraqis couldn't shoot—at all."
46. Keefe 2005 interview.
47. *March Up*, 92.
48. Ibid., 95–96.
49. Ibid., 92.
50. Ibid., 172.
51. Eric Mitchell, "A Critical Vulnerability," *Marine Corps Gazette* (March 2014), https:// www.mca-marines.org/gazette/2014/03/critical-vulnerability
52. Gordon, *Cobra II*, 300; West, *March Up*, 77.
53. West, *March Up*, 171. For a detailed analysis before the war of the need for the Dragon Eye, see James McCain, "'Dragon Eye': The Small Unit Commander's Personal Unmanned Aerial Vehicle," *Marine Corps Gazette* (January 2001), https://www.mca-marines.org/gazette/2001/01/dragon-eye-small-unit-commanders-personal-unmanned-aerial-vehicle: "It is my opinion that the answer is that warfighters in the Marine Corps do not have timely access to any of the other current UAV systems that are operational, either Marine Corps operated systems or systems operated by other Services.... What does all this mean to the Marine warfighter? Simply stated, it means that at the battalion level and below, Marine warfighters do not have access to the same current intelligence provided by UAVs that is available to higher headquarters. It means that the vast majority of UAV intelligence that is provided to the battalion commander is a day late and a dollar short. It also means that there is a UAV void that needs to be filled within the Marine Corps. So enter Dragon Eye."
54. Buhle, Iraq War Oral Histories; "Rifle Combat Optic" on United States Marine Corps website, http://www.marines.com/operating-forces/equipment/gear/rifle-combat-optic (accessed April 20, 2014). The website quotes General James N. Mattis: "The ACOG mounted on the M-16 service rifle has proven to be the biggest improvement in lethality for the Marine infantryman since the introduction of the M! Garand (the first semi-automatic rifle used in combat) in World War II."
55. West, *March Up*, 130; Mick Tucker, *Among Warriors in Iraq: True Grit, Special Ops, and Raiding in Mosul and Fallujah* (Guilford, CT: Globe Pequot Press, 2005), 112–13.
56. Sara Wirtala Bock, "Marine Corps Optics: Improving the 'Sight' of the Individual Marine," *Leatherneck* (April 2009), https://www.mca-marines.org/leatherneck/2009/04/marine-corps-optics-improving-sight-individual-marine (accessed April 20, 2014).
57. Michael Hanson, "Coin Perspectives," *Marine Corps Gazette* (April 2009), https://www.mca-marines.org/gazette/2009/04/coin-perspectives (accessed May 2, 2014).
58. Conway quoted in Gary Livingston, *An Nasiriyah: The Fight for the Bridges* (North Topsail Island, NC: Caison Press, 2003), 14.
59. Email communication from Colin Keefe to author, March 6, 2014.
60. Keefe 2005 interview.
61. Email communication from Keefe to author, March 6, 2014.
62. Keefe 2005 interview.
63. Ibid.
64. Ibid.
65. Ibid.

CHAPTER 2: **Of War Porn and Pleasure in Killing**

1. Michael Walzer, *Just and Unjust Wars: A Moral Argument with Historical Illustrations* (New York: Basic Books, 1977), 311. See also Robert Westbrook, "'I Want a Girl, Just like the Girl

That Married Harry James': American Women and the Problem of Political Obligation in World War II," *American Quarterly* 42 (1990): 587–614.

2. William James, "The Moral Equivalent of War," *McClures Magazine*, August 1910, reprinted in *The Writings of William James*, ed. John J. McDermott (New York: Random House, 1967), 660–71 (emphasis in original).

3. The indexical quality of still and motion pictures was enunciated most clearly by film theorist Andre Bazin, *What Is Cinema?*, trans. Hugh Gray (1946–57; repr., Berkeley: University of California Press, 2004), 1:13; Steven Shaviro, "Emotion Capture: Affect in Digital Film," *Projections* 1 (Winter 2007): 63–82.

4. Vivian Sobchak, *Carnal Thoughts: Embodiment and Moving Image Culture* (Berkeley: University of California Press, 2004), 142. Sobchak writes: "The moving picture is a visible representation not of activity finished or past but of activity coming into being and being. Furthermore, and even more significant, the moving picture not only visibly represents moving objects but also—and simultaneously—presents *the very movement of vision itself*." Very helpful in distinguishing between the cognitive and emotional effect of, on the one hand, still images, and, on the other, moving pictures, is Susan Sontag, "Looking at War: Photography's View of Devastation and Death," *New Yorker*, December 9, 2002, reprinted in Sontag, *Regarding the Pain of Others* (New York: Farrar, Straus, and Giroux, 2004), 18–94. Sontag wrote: "Non-stop imagery (television, streaming video, movies) surrounds us, but when it comes to remembering, the photograph has the deeper bite. Memory freeze frames; its basic unit is the single image. In an era of information overload, the photograph provides a quick way of apprehending something and a compact form of memorizing it. The photograph is like a quotation, or a maxim or proverb" (22).

5. World War II GIs quoted in Paul Virilio, *War and Cinema: The Logistics of Perception* (London: Verso, 1989), 48; Philip Caputo, *A Rumor of War* (New York: Henry Holt, 1977), 306; Sergeant Connors quoted in *Shootout: D-Day Fallujah*, The History Channel, 2006, accessed on YouTube, November 19, 2012, http://www.youtube.com/watch?v=7cx4bEJVKe8&list=PLRrm Cq9BB1FBolfVjmbV-A2eikaGDGoQj&index=38; Private Stokes quoted in Patrick K. O'Donnell, *We Were One: Shoulder to Shoulder with the Marines Who Took Fallujah* (Cambridge, MA: De Capo Press, 2006), 77.

6. Lawrence M. Ashman, "The Corps' Film Archives," *Marine Corps Gazette* (April 1951): 48.

7. Brian Thompson and Brook R. Kelsey, "Combat Camera in the Long War," *Marine Corps Gazette* (April 2010), http://pqasb.pqarchiver.com/mca-members/access/2010273701.html?FMT =FT&FMTS=ABS:FT:TG:PAGE&type=current&date=Apr+2010&author=Brian+Thompson %3BBrook+R+Kelsey&pub=Marine+Corps+Gazette&edition=&startpage=24&desc=Combat +Camera+in+the+Long+War (accessed January 27, 2013).

8. Andrew Lubin, "The New 'Strategic' Combat Correspondent in Afghanistan," *Leatherneck* (September 2011), http://pqasb.pqarchiver.com/mca-members/access/2443835371.html?FMT =FT&FMTS=ABS:FT:PAGE&type=current&date=Sep+2011&author=Andrew+Lubin& pub=Leatherneck&edition=&startpage=48&desc=The+New+%22Strategic%22+Combat+ Correspondent+in+Afghanistan (accessed January 27, 2013).

9. Lawrence H. Suid, "Hollywood, the Marines, and the Korean War," *Marine Corps Gazette* (March 2002): 42. See also Lawrence H. Suid, "The Image of the Marines and John Wayne," in *Guts and Glory: The Making of the American Military Imagine in Film* (Lexington: University of Kentucky Press, 2002); Clayton R. Koppes and Gregory D. Black, *Hollywood Goes to War: How Politics, Profits and Propaganda Shaped World War II Movies* (Berkeley: University of California Press, 1987); Eckhart quoted in "Battle L.A. Seizes the Day with U.S. Marines Product Placement," PlaceVine: The Brand Integration Service, Blog, http://archive.placevine.com/blog

/2011/03/15/battle-la-seizes-the-day-with-u-s-marines-product-placement/ (accessed March 27, 2011).

10. Dana Calvo, "Theaters Pull Plug on 'Enduring Freedom,'" *Los Angeles Times*, October 18, 2002, http://articles.latimes.com/print/2002/0ct/18/entertainment/er-calvo18 (accessed June 25, 2012); Tony Perry, "Military Short May Hit Theaters," *Los Angeles Times*, November 17, 2003, http://articles.latimes.com/print2003/nov/17/entertainment/et-perry (accessed June 25, 2012).

11. George Zornick, "The Porn of War," *The Nation*, October 10, 2005, http://www.thenation.com/article/porn-war (accessed June 4, 2012); Louise Roug, "Extreme Cinema Verite," *Los Angeles Times*, March 14, 2005, http://articles.latimes.com/2005/mar/14/world/fg-videos14 (accessed October 15, 2012); Mark Glaser, "YouTube Offers Soldier's Eye View of Iraq War," *Mediashift*, January 25, 2006, http://www.pbs.org/mediashift/2006/01/digging_deeperyoutube_offers_s_1.html (accessed April 17, 2007); Kari Anden-Papadopoulos, "US Soldiers Imaging the War on YouTube," *Popular Communication* 7 (2009): 17–27; Christian Christensen, "Uploading Dissonance: YouTube and the US Occupation of Iraq," *Media, War and Conflict* 2 (2008): 155–75; Jennifer Terry, "Killer/Entertainments," *Vectors: Journal of Culture and Technology in a Dynamic Vernacular* 5 (2007), http://vectors.usc.edu/issues/5/killerentertainments/ (accessed November 14, 2012).

12. With this statement I am, for the sake of clarity of argument and subsequent usage, shortening the history of war pornography. In the broadest sense, war pornography—the representational aggrandizement of war—has accompanied the human institution of armed conflict from its birth. Recently, the term "war porn" was used by Jean Baudrillard to describe the mostly still images of U.S. soldiers' abuse of Iraqi prisoners taken at Abu Ghraib prison Iraq in 2003 and 2004. Jean Baudrillard, "War Porn," *Journal of Visual Culture* 5 (2006): 86–88. For incisive essays on visual culture of war and that touch on war porn, see Kari Anden-Papadopoulos, "Body Horror on the Internet: US Soldiers Recording the War in Iraq and Afghanistan," *Media, Culture and Society* 31 (2009): 921–38; Matteo Pasquinelli, "Warporn, Warpunk! Autonomous Videopoesis in Wartime," *Sarai Reader* (2005), http://www.globalresearch.ca/it-s-called-war-porn/ (accessed October 3, 2012); David Swanson, "The Iraq War as a Trophy Photo," *TomDispatch.com*, http://www.tomdispatch.com/index.mhtml?pid=91318 (accessed April 18, 2007).

13. See, for example, Riz Khan, "The 'YouTube Wars,'" *Aljazeera*, June 16, 2010, http://www.aljazeera.com/programmes/rizkhan/2010/06/20106157235198947.html (accessed September 29, 2012). The integration of war with viral video and social media only continues, of course. See, for example, Scott Shane and Ben Hubbard, "Isis Displaying a Deft Command of Varied Media," *New York Times*, August 31, 2014, 1.

14. Singer quoted in Jessica Ramirez, "Carnage.com," *Daily Beast*, May 9, 2010. http://www/thedailybeast.com/newsweek/2010/04/30/carnage-com (accessed March 7, 2012).

15. Steve Mraz, "'Proud to Be an American' with a Slightly Harder Tune," *Stars and Stripes*, July 18, 2005, http://stripes.com/article.asp?section=104&article=29620&archive=true (accessed April 18, 2007).

16. Stacey Peebles, *Welcome to the Suck: Narrating the American Soldier's Experience in Iraq* (Ithaca: Cornell University Press, 2011), 48.

17. In a representative feminist definition of pornography, psychologist Diana E. H. Russell writes, "It is sexually explicit material that represents or describes degrading or abusive sexual behavior so as to endorse and/or recommend the behavior as described." Russell, "Pornography and Rape: A Causal Model," *Political Psychology* 9 (1988): 42. For valuable overview and primary source material on pornography, see *Feminism and Pornography*, ed. Drucilla Cornell (New York: Oxford University Press, 2000).

18. Robin Morgan, *Going Too Far: The Personal Chronicle of a Feminist* (New York: Random House, 1978), 133.

19. Among the many works chronicling the pleasure principle of war, see J. Glenn Gray, *The Warriors: Reflections on Men in Battle* (New York: Harper, 1967); Joanne Bourke, *An Intimate History of Killing: Face-to-Face Killing in 20th-Century Warfare* (New York: Basic Books, 1999). I discuss the distinct post–Civil War rise of U.S. cultural representations of pleasure in fighting in *Brutes in Suits: Male Sensibility in America, 1890–1920* (Baltimore: Johns Hopkins University Press, 2007), 197–268.

20. Evan Wright, *Generation Kill: Devil Dogs, Iceman, Captain America and the New Face of American War* (New York: Putnam, 2004), 336.

21. Nathaniel Fick, *One Bullet Away: The Making of a Marine Officer* (Boston: Houghton Mifflin, 2005), 216.

22. Ibid., 261.

23. Mattis quoted in John R. Guardino, "Breaking the Warrior Code," *American Spectator*, February 11, 2005, http://spectator.org/archives/2005/02/11/breaking-the-warrior-code/ (accessed December 27, 2012).

24. Craig M. Cameron, *American Samurai: Myth, Imagination, and the Conduct of Battle in the First Marine Division, 1941–1951* (New York: Cambridge University Press, 1994), 45; Dave Grossman, *On Killing: The Psychological Cost of Learning to Kill in War and Society* (New York: Little, Brown, 1995), 314, 315; Bourke, *Intimate History of Killing*, 10–11.

25. Jennifer Terry, "Killer Entertainments," *Vectors: Journal of Culture and Technology in a Dynamic Vernacular* 5 (2007), http://vectors.usc.edu/issues/5/killerentertainments/ (accessed September 5, 2014).

26. Thomas Schatz, "Old War/New War: *Band of Brothers* and the Revival of the WWII Film," *Film and History* 32 (2002): 75, as quoted by J. David Slocum, general introduction to *Hollywood and War*, ed. J. David Slocum (New York: Routledge, 2006), 2.

27. Caputo, *Rumor of War*, 14; Michael Herr, *Dispatches* (New York: Alfred A. Knopf, 1978), 195–96.

28. Jeanine Basinger, *The World War II Combat Film: Anatomy of a Genre* (New York: Columbia University Press, 1986). See also Suid, *Guts and Glory*, 42–135; Thomas Schatz, "World War II and the Hollywood 'War Film,'" in Slocum, *Hollywood and War*, 147–56; Thomas Doherty, *Projections of War: Hollywood, American Culture, and World War II* (New York: Columbia University Press, 1993); John Bodnar, *The Good War in American Memory* (Baltimore: Johns Hopkins University Press, 2011), 130–65.

29. Bosley Crowther, review of *Wake Island*, *New York Times*, September 2, 1942, http://movies.nytimes.com/movie/review (accessed July 15, 2009).

30. Bosley Crowther, review of *Gung Ho!*, *New York Times*, June 26, 1944, http://movies.nytimes.com/movie/review (accessed July 15, 2009).

31. Cameron, *American Samurai*, 259–62; Bassinger, *World War II Combat Film*, 151–53.

32. Caputo, *Rumor of War*, 14; William Broyles Jr., "Why Men Love War," *Esquire*, November 1984, 56.

33. Anthony Swofford, *Jarhead* (New York: Scribner, 2003), 5, 6–7.

34. "Iraq: Bird Is the Word," YouTube video, uploaded by Jacob J. Tanenbaum, August 20, 2006, http://www.youtube.com/watch?v=VBmhA6z6ieY&list=PLRrmCq9BB1FBolfVjmbV-A2eikaGDGoQj&index=13; Tanenbaum explains his intent in response to a comment in December 2011.

35. See, for example, comment by petehjr1, YouTube, posted April 2011.

36. Anden-Papadopoulos, "Body Horror on the Internet," 923. Also helpful for the context to U.S. Defense Department presence on YouTube, Cori E. Dauber, *YouTube War: Fighting*

in a World of Cameras in Every Cell Phone and Photoshop on Every Computer, e-book (Carlisle Barracks, PA: Strategic Studies Institute, 2009).

37. That digital cameras and editing software accompanied U.S. marines and soldiers in Iraq is a central fact of that war's social history. One statement of the point comes from U.S. Marine Sergeant Jason Shaw in MTV's documentary aired on July 20, 2006; the accompanying article can be accessed at http://www.mtv.com/news/articles/1536780/iraq-uploaded-soldiers-post-war-footage-online.jhtml. See also, Edward Wyatt, "Now on YouTube: Iraq Videos of U.S. Troops under Attack," *New York Times*, October 6, 2006, A1, 21.

38. In terms of the hard-core and soft core porn typologies, see, for example, "War Porn, a Guide," by NewsweekVideo, YouTube, uploaded June 17, 2010, http://www.youtube.com/watch?v=1VeyX_gcrNE&list=PLRrmCq9BB1FBolfVjmbV-A2eikaGDGoQj&index=2.

39. "3rd Battalion, 1st Marines in Fallujah Iraq," YouTube video, uploaded by swede1775, November 30, 2006, http://www.youtube.com/watch?v=SW9bBUedzWI&list=PLRrmCq9BB1FBolfVjmbV-A2eikaGDGoQj&index=32; MarineCorpsMarksman, commenting on "3rd Battalion, 1st Marines," 2009.

40. Jegor666, commenting on "3rd Battalion, 1st Marines," 2007; DScherer09, commenting on "3rd Battalion, 1st Marines," 2008; 75ranger101, commenting on "3rd Battalion, 1st Marines," 2010.

41. leandean100, commenting on "3rd Battalion, 1st Marines," 2013; dmarnott, commenting on "3rd Battalion, 1st Marines," 2011; USMCLP, commenting on "3rd Battalion, 1st Marines," 2012.

42. For example, "YEA it is shown in bootcamp." mxmarinekid08, commenting on "3rd Battalion, 1st Marines," 2008.

43. "Videos in boot camp now? No more iron sights, rolled sleeves, the whole shooting badge thing, or sleeve tattoos (personal preference). What is my Corps coming too. Eh fuck it. We're still the most badass fighting force the world has ever seen and we're gentlemen to boot! Semper Fucking Fidelis Oooh Rahh Yut Yut Kill!" stoker20010543, commenting on "33rd Battalion, 1st Marines," 2008.

44. "Cool, i wish i could play there too. Is there a way to change sides when you come to Germany next time? America is a great land." magkeinepolizisten, commenting on "3rd Battalion, 1st Marines," 2008.

45. Bobbias commenting on "3rd Battalion, 1st Marines," 2008.

46. "Predator Drone," uploaded by fogarty23, June 29, 2011, http://www.youtube.com/watch?v=Ks8nOvPSJfY&list=PLRrmCq9BB1FBolfVjmbV-A2eikaGDGoQj&index=4. freechristiana, commenting on "Predator Drone," 2012.

47. "Iraq—Throwing a Grenade in House," YouTube video, uploaded by Dru Mat, October 1, 2010, http://www.youtube.com/watch?v=3_eCW_iXCps&list=PLRrmCq9BB1FBolfVjmbV-A2eikaGDGoQj&index=30. Bee Burgers, commenting on "Iraq—Throwing Grenade in House," 2012.

48. Redhawk092, commenting in 2008 on "Marine Get Some: Die MF Die," YouTube video, uploaded by MarineTroopSupport, August 12, 2007, http://www.youtube.com/watch?v=5JXSz2aNefA.

49. miked90123, commenting on "Iraq: Bird Is the Word," 2012.

50. For Marines' view of Japanese in World War II, see Cameron, "Images of the Japanese 'Other' Defined: Guadalcanal and Beyond," in *American Samurai*, 89–129. For the crucial paradigm of dehumanized, racialized other, John W. Dower, *War without Mercy: Race and Power in the Pacific War* (New York: Pantheon Books, 1986).

51. southernpipeman, commenting on "Iraq: Bird Is the Word," 2011.

52. The communicative purpose of the videos showing the 2014 ISIS beheadings of Amer-

ican journalists James Foley and Steven Sotloff may very well be multifaceted, and their meanings many. The stated purpose was deterrence of further U.S. attacks by air of ISIS military forces. But the audacity of the recorded beheadings could also work to pull more fighters into the ISIS fold, thereby paralleling the function of American YouTube war porn. For one of many discussions of the videos, David Carr, "With Videos of Killings, ISIS Sends Medieval Message by Modern Method," *New York Times*, September 8, 2014, B1.

53. heelhook86, commenting on "Marine Get Some," 2008; aggiecornhusker, commenting on "Marine Get Some," 2008.

54. "Bullet with a Name on It," YouTube video, uploaded by WilcoUSMC on March 19, 2006. rgelhart, commenting on "Bullet with a Name on It," 2006; Marine137, commenting on "Bullet with a Name on It," 2006.

55. Several comments during 2006 suggest they were made by deployed marines. For instance, "Semper Fi! you guys have no idea what kind of moral boost this does for us overseas. god bless and thanks." USMC0331, commenting on "Bullet with a Name on It," 2006.

56. "Iraq War—Let the Bodies Hit the Floor," YouTube video, uploaded by dcg9 on July 27, 2009, http://www.youtube.com/watch?v=91RBb8mCejc&list=PLRrmCq9BB1FBolfVjmbV-A2eikaGDGoQj&index=12. Report of Drowning Pool's song played by Coalition Psychological Operations, O'Donnell, *We Were One*, 100.

57. TheCarol001, commenting on "Let the Bodies Hit the Floor," 2011; turkzil commenting on "Let the Bodies Hit the Floor," 2012.

58. faithandrace, commenting on "Bird Is the Word," 2011; spookje327, commenting on "Bird Is the Word," 2012.

59. "U.S. Marine Kills Iraqi," YouTube video, uploaded by randytune88, on July 26, 2009. Erodacly Ace, commenting on "U.S. Marine Kills Iraqi," 2012; bammywammy1997 commenting on "Bird Is the Word," 2011.

60. morelltechyahoocom, commenting on "Bullet with a Name on It," 2009; TxBoarHunter, commenting on "Bullet with a Name on It," 2008.

61. beefertown, commenting on "U.S Marine Kills Iraqi," 2012; miked90123, commenting on "Bird Is the Word," 2012.

62. lostinthegame77, commenting on "Bullet with a Name on It," 2009; Northernsea, commenting on "Bullet with a Name on It," 2006.

63. Roger Stahl, *Militainment, Inc.: War, Media and Popular Culture* (New York: Routledge, 2010). Stahl's definition of *militainment* is taken from his documentary film, *Militainment, Inc.: Militarism and Popular Culture* (Northampton, MA: Media Education Foundation, 2007), streaming at rogerstahl.info.

64. Stacy Takacs, *Terrorism TV: Popular Entertainment in Post-9/11 America* (Lawrence: University Press of Kansas, 2012), 165.

65. *Over There*, FX, July 27, 2005–October 26, 2006.

66. Graeme Turner and Jinna Tay, eds., *Television Studies after TV: Understanding Television in the Post-Broadcast Era* (New York: Routledge, 2009), 2.

67. *War Stories with Oliver North*, Fox News, http://www.foxnews.com/on-air/war-stories/index.html (accessed November 10, 2012).

68. *Deadliest Warrior*, Spike TV, April 7, 2009–September 14, 2011.

69. Wright, *Generation Kill*, 5.

70. Ibid., 6–7.

71. *Generation Kill*, episode 3, DVD (Santa Monica, CA: HBO, 2008).

72. "For God and Country: A Marine Sniper's Story," MSNBC, aired October 17, 2009. Transcript at http://www.livedash.com/transcript/for_god_and_country__a_marine_sniper's_story/5304/MSNBC/Saturday_October_17_2009/97777/.

Notes to Pages 62–69   187

73. "Shootout: D-Day Fallujah," The History Channel, season 1, episode 1, August 2005.
74. https://www.youtube.com/watch?v=g3zZmOR0jYc (accessed November 18, 2012).
75. Chris Sy, commenting on "Shootout: D-Day Fallujah," YouTube video, http://www.youtube.com/watch?v=7cx4bEJVKe8&list=PLRrmCq9BB1FBolfVjmbV-A2eikaGDGoQj&index=38, 2012.
76. Japai, commenting on "Bullet with a Name on It," 2006.

CHAPTER 3: Fallujah, *First to Fight,* and Ludology

1. Kevin Sites, "What Happened In the Fallujah Mosque," MSNBC.com, May 4, 2005 (accessed February 15, 2006).
2. "Developer Addresses Fallujah Game Cancellation," *GamePolitics.com,* May 1, 2009, http://www.gamepolitics.com/2009/05/01/developer-addresses-fallujah (accessed May 4, 2010); Dan Ephron, "The Battle over the Battle of Fallujah," *Newsweek,* June 6, 2009, http://www.newsweek.com/id/200861.
3. Author interview with Peter Tamte, chief executive officer of the Destineer Corporation, June 20, 2008.
4. "Marine Corps Family of Tactical Decisionmaking Simulations," *Marine Corps Gazette* (September 2004), http://pqasb.parchiver.com/mca-members/access/Marine+Corps+Family+of+Tactical+Decisionmaking+Simulations (accessed January 2, 2013).
5. Tamte interview.
6. Ephron, "The Battle over the Battle of Fallujah."
7. Johan Huizinga, *Homo Ludens* (1938; repr., Boston: Beacon Press, 1955), 89.
8. For the close relationship between war and sport and play in American culture, see my *Brutes in Suits: Male Sensibility in America, 1890–1920* (Baltimore: Johns Hopkins University Press, 2007), 160–78, 242–54.
9. For background on the battle for Fallujah, Bing West, *No True Glory: A Frontline Account of the Battle for Fallujah* (New York: Bantam Books, 2005); Thomas E. Ricks, *Fiasco: The American Military Adventure in Iraq* (New York: Penguin Books, 2006), 398–403.
10. Shupp quoted in Patrick K. O'Donnell, *We Were One: Shoulder to Shoulder with the Marines Who Took Fallujah* (New York: De Capo Press, 2006), 61–62.
11. "Red Cross Estimates 800 Iraqi Civilians Killed in Fallujah," Democracy Now.org, November 17, 2004, http://www.democracynow.org/2004/11/17/red_cross_estimates_800_iraqi_civilians (accessed February 24, 2013).
12. Jim Krane, "U.S. Aircraft Stacked Up over Fallujah Targets," *Philadelphia Daily News,* November 12, 2004, 30. "No fewer than 20 types of aircraft have been thrown into the fight, including 10 fixed-wing planes, three types of helicopters and seven kinds of unmanned drones."
13. West, *No True Glory,* 260–63.
14. Earl J. Catagnus Jr., Brad Z. Edison, James D. Keeling, and David A. Moon, "Infantry Squad Tactics: Some of the Lessons Learned during MOUT in the Battle for Fallujah," *Marine Corps Gazette* (September 2005): 80–89.
15. Paul McLeary, "Bing West on the Marines, the Press, and Iraq," *Columbia Journalism Review Daily,* November 4, 2005, http://www.cjrdaily.org/the_water_cooler/bing_west_on_the_marines (accessed April 19, 2006). Thomas Ricks has described Fallujah as "probably the toughest battle the U.S. military had seen since the Vietnam War." *Fiasco,* 399.
16. West, *No True Glory,* 7.
17. O'Donnell, *We Were One,* 76–77.
18. "3rd Battalion, 1st Marines in Fallujah Iraq," YouTube video, uploaded by swede1775, November 30, 2006, http://www.youtube.com/watch?v=SW9bBUedzWI&list=PLRrmCq9BB1FBolfVjmbV-A2eikaGDGoQj&index=32.

19. "Holy War: Evangelical Marines Prepare to Battle Barbarians," *Common Dreams Newscenter*, November 7, 2004, http://www.commondreams.org/ (accessed February 10, 2005).

20. The presence of such video games is a commonplace in the social history of marines in camp in 2004 and beyond—that is, those deployed after the initial invasion in 2003. The point is confirmed in my interview with Mike Ergo (sergeant USMC ret.), January 25, 2013.

21. For intelligent critical summary and assessment of these categories, see Alexander R. Galloway, *Gaming: Essays on Algorithmic Culture* (Minneapolis: University of Minnesota Press, 2006).

22. *Brown v. Entertainment Merchants Association*, 131 S. Ct. 2729 (2011).

23. My understanding of the *Brown* decision benefited greatly from Robert Bryan Norris Jr., "It's All Fun and Games until Someone Gets Hurt: *Brown v. Entertainment Merchants Association* and the Problem of Interactivity," *North Carolina Journal of Law and Technology* 13 (2011): 81–116.

24. *Brown v. Entertainment Merchants*, 2767.

25. Interdisciplinary scholarship on the military entertainment complex is considerable and impressive. Leading works include Tim Lenoir, "All but War Is Simulation: The Military-Entertainment Complex," *Configurations* 8 (Fall 2000): 289–335; James der Derian, *Virtuous War: Mapping the Military-Industrial-Media-Entertainment Network* (Boulder: Westview Press, 2001); Ed Halter, *From Sun Tzu to Xbox: War and Video Games* (New York: Thunder Mouth Press, 2006); Roger Stahl, *War, Media, and Popular Culture* (New York: Routledge, 2010).

26. Macedonia quoted in Heather Chaplin and Aaron Ruby, *Smartbomb: The Quest for Art, Entertainment, and Big Bucks in the Videogame Revolution* (Chapel Hill: Algonquin Books, 2006), 200–201. Lenoir speaks of *Ender's Game*'s influence on military simulation, including it being required reading at the Marine University in Quantico, Virginia. "All But War Is Simulation," 305.

27. Orson Scott Card, *Ender's Game* (1977; New York: Starscape, 1985), 258–59.

28. Ibid., 296–97.

29. Ibid., 297–98.

30. Ibid., 298, 274, 298.

31. 1979 print advertisement by Midway Manufacturing Company, http://www.marvin3m.com/arcade/msubmar.htm (accessed June 11, 2007).

32. For Rotberg quote and the Army's collaboration with Atari, see Halter, *From Sun Tzu to Xbox*, 131–32.

33. Ronald Reagan, "Remarks during a Visit to Walt Disney World's Epcot Center," March 8, 1983, http://www.presidency.ucsb.edu/ws/index.php?pid=41022 (accessed February 24, 2013).

34. *Modeling and Simulation: Linking Entertainment and Defense* (Washington, DC: National Academy Press, 1997), viii.

35. Ibid., 39.

36. USC Institute for Creative Technologies (ICT) website, http://ict.usc.edu/about (accessed June 5, 2008).

37. "Marines Using ICT Flatworld Technologies," ICT website, April 5, 2007, http://ict.usc.edu/news/item/marine_corps_using_ict_flatworld_technologies (accessed June 5, 2008).

38. Pete Muller, "The Infantry Immersion Trainer," *Marine Corps Gazette* (September 2008): 14, 16, 18; Dennis Thompson, "MAGTF Simulations," *Marine Corps Gazette* (September 2009): 26–29.

39. Rob Ridell, "Doom Goes to War," *Wired*, April 1997, http://www.wired.com/wired/archive/5.04/ff_doom.html (accessed December 16, 2006). The Marines' embrace of video games is also covered in Lenoir, "All But War Is Simulation," 19–20; Halter, *Sun Tzu to Xbox*, 165–69.

40. Marine Corps Order 1500.55, April 12, 1997, http://www.usmc.mil/news/publications/Documents/MCO%201500.55.pdf (accessed June 24, 2007).
41. N. J. Jernigan, "Marine Doom," *Marine Corps Gazette* (August 1997): 19–20 (accessed July 6, 2012).
42. Halter, *Sun Tzu to Xbox*, 168.
43. "Next Generation of Close Combat Games on the Horizon," *Destineer*, April 1, 2004, http://www.firsttofight.com/html (accessed March 29, 2006).
44. *Close Combat: First to Fight* (Destineer Corporation, 2005).
45. Ibid.
46. Interview with Tamte.
47. Particularly helpful in discussing Hollywood's influence on combat video gaming is Halter, *Sun Tzu to Xbox*, 272–90.
48. Author interview with David Kuykendall, game designer, Destineer Studios, June 13, 2008.
49. Ibid.
50. Interview with Tamte.
51. *First to Fight* Instruction Manual.
52. Keykendall emphasized this point in his interview. Note, however, in video gaming's rapid cultural ascent, considerable capital and creative investment are being made to match film in representing facial expression. See, for instance, Sharon Waxman, "Cyberface," *New York Times*, October 15, 2006, sec. 2, 1, 14; Tom Bissell, "Voicebox 360: The Queen of Video-Game Acting," *New Yorker*, August 15 and 22, 2011, 48–53.
53. Waxman, "Cyberface."
54. Tamte interview.
55. Kuykendall interview.
56. David B. Nieborg, "Mods, Nay! Tournaments, Yay!—The Appropriation of Contemporary Game Culture by the U.S. Military," *fibreculture* 8 (2008), http://journal.fibreculture.org/issue8 (accessed January 29, 2009).
57. Richard Siklos, "Not in the Real World Anymore: Upstaged by MySpace, MTV Tries Virtual Reality," *New York Times*, September 18, 2006, sec. C, 1, 6; Josh Quitner, "A PR-Rated Second Life," *Time*, January 19, 2009, 52; Jay Hart, "For the Second Year in a Row, Video Creates a Pocono Star," *Allentown Morning Call*, June 12, 2006, sec. C, 1, 3.
58. Sue Halpern, "Virtual Iraq," *New Yorker*, May 19, 2008, 32–37; Dan Frosch, "Fighting the Terror of Battles That Rage in Soldiers' Heads," *New York Times*, May 13, 2007, http://www.nytimes.com/2007/05/13/us/13carson.html (accessed May 20, 2007).
59. Ellen Manaker, Susan Coleman, Joe Collins, and Marci Murawski, "Harnessing Experiential Learning Theory to Achieve Warfighting Excellence," *I/ITSEC* 2006, 4.
60. Macedonia quoted in Halter, *Sun Tzu to Xbox*, 201.
61. Jeremy Saucier uses the term *virtual veteran* in his unpublished essay, "'Empower Yourself, Defend Freedom': *America's Army*, the Army Experience Center, and the Rise of the Virtual Veteran in Recent America," in author's possession.
62. *Modeling and Simulation*, 28.
63. Halter, *From Sun Tzu to Xbox*, 161.
64. Note, in many discussions of the interactive appeal of combat video games, the word and quality of *realism* are often invoked in praise of particular games. See, for example, Bennet Ring, "Call of Duty 4 AU Interview," *IGN*, July 22, 2007, http://www.ign.com/articles/2007/07/23/call-of-duty-4-au-interview (accessed September 8, 2014). In this article, retired Army lieutenant colonel Hank Keirsey, military advisor for *Call of Duty* 4, comments: "These games you end up going from one extreme fire-fight to the next most extreme situation to the

next most extreme situation. The realism is absolutely real if you're the one in thousand guys that ends up in that particularly bad sector, and got ambushed. Absolutely real."

65. Taylor quoted in Amy Harmon, "More than Just a Game, but How Close to Reality?," *New York Times*, April 3, 2003, http://www.iraqwararchive.org/data/apr03/US/nyt11.pdf (accessed June 9, 2009).

66. Macedonia quoted in Halter, *From Sun Tzu to Xbox*, 201.

67. Jose Antonio Vargas, "Virtual Reality Prepares Soldiers for Real War," *Washington Post*, February 14, 2006, http://www.washingtonpost.com/wp-dyn/content/article/2006/02/13/AR2006021302437.html.

68. Karen Till, "Memory Studies," *History Workshop Journal* 62 (Autumn 2006): 333.

69. In considering this comparison between video gaming and football and their function to warfighting, journalist Richard Harding Davis's observation of former college football players fighting in the Spanish-American War is edifying: "That same spirit that once sent these men down a white-washed field against their opponents rush-line was the spirit that sent [them] through the hot grass at Guasimas, not shouting, as their friends the cowboys did, but each with his mouth tightly shut, with his eyes on the ball, and moving in obedience with the captain's orders." Richard Harding Davis, *The Cuban and Porto Rican Campaigns* (London: William Heinemann, 1899), 140–41. For discussion of this comparison and its implications, see Pettegrew, *Brutes in Suits*, 242–55.

70. *Modeling and Simulation*, 1, 14.

71. James Dao, "To Prepare for War, G.I.'s Get a Dress Rehearsal," *New York Times*, November 29, 2004, sec. N 22, 24.

72. Mark Bowden, *Black Hawk Down: A Story of Modern War* (New York: Signet Books, 1999).

73. Statement of John Rhodes, Commanding General, USMC Combat Development Command, before the Senate Armed Services Committee, on October 20, 1999, concerning Marine Corps Experimentation Efforts, 1, http://www.fas.org/man/congress/1999/99-10-19rhodes.htm. (accessed April 10, 2010).

74. Ibid., 3, 5.

75. *Urban Warrior: Conceptual Experimental Framework* (Quantico, VA: Marine Warfighting Laboratory, April 1998).

76. General Charles C. Krulak, "The Strategic Corporal: Leadership in the Three Block War," *Marines Magazine*, January 1999, http://www.au.af.mil/au/awc/awcgate/usmc/strategic_corporal.htm. (accessed April 20, 2010).

77. Ibid.

78. Tamte, "Developer Diary #1," *Gamespy*, http://xbox.gamespy.com/xbox/close-combat-first-to-fight/579732p1.html (accessed October 11, 2011); *First to Fight* Instruction Manual, 7.

79. Ibid.

80. Kuykendall interview.

81. Email from Tamte to author, May 23, 2010.

82. Kuykendall interview.

83. *First to Fight*, Instruction Manual, 28.

84. *Urban Warrior, Conceptual Experimental Framework* (Quantico, VA: MAGTF Warfighting Center, 1998).

85. *First to Fight*, Instruction Manual, 22.

86. Tamte, "Developer Diary #6," *Gamespy*, http://xbox.gamespy.com/xbox/close-combat-first-to-fight/579732p1.html (accessed September 23, 2011).

87. Both comments from *IGN* Discussion Board, http://boards.ign.com/close_combat, WWC33, posted April 9, 2005; radar9999, posted November 26, 2005.

88. Kuykendall interview.

89. Tamte interview. Researchers at the Office of Naval Research reported in 2010 that "video game players perform 10 to 20 percent higher in terms of perceptual and cognitive ability than people that are non-game players." Bob Freeman, "Researchers Examine Video Game Benefits," United States Department of Defense, News, American Forces Press Service, http://www.defense.gov/news/newsarticle.aspx?id=57695 (accessed July 26, 2010).

90. *IGN* Discussion Board, Owen36, posted April 16, 2006; Lurkers, posted April 26, 2006.

91. *First to Fight* Instruction Manual, 23.

92. *The U.S. Army/Marine Corps Counterinsurgency Field Manual* (2006; Chicago: University of Chicago Press, 2007), 47–40. The term "paradigm shattering" is from Sarah Sewall's introduction to the volume, "A Radical Field Manual," xxxv.

93. Interview with Mike Ergo, January 25, 2013.

94. "Developer Addresses Fallujah Game Cancellation," GamePolitics.com, May 1, 2009, http://www.gamepolitics.com/2009/05/01/developer-addresses-fallujah (accessed May 22, 2009); Dan Ephron, "The Battle over the Battle of Fallujah," *Newsweek*, June 6, 2009, http://www.newsweek.com/id/200861/ (accessed June 24, 2009).

95. Evan Wright, "Dazed and Confused," *Variety* 400 (December 2005), 60 (accessed March 30, 2007).

CHAPTER 4: Counterinsurgency and "Turning Off the Killing Switch"

1. Commanding Officer P. J. Malay, "Letter of Appreciation," June 23, 2005, 3D Battalion, Fifth Marines, 3/5 Battalion File, United States Marine Corps Archives, Quantico, Virginia.

2. Interview of Corporal Nick Poletis, December 12, 2003, United States Marine Corps History Division, Oral History Interview, Camp Lejeune, North Carolina. In the same Camp Lejeune interview, 3/2 sergeant Mark Delarosa emphasized, "We were helping people out" after fighting in Nasiriyah: "I wish people at home would know that."

3. Author interview with Thomas Lowry, August 8, 2012.

4. General Mattis quotes are taken from Donovan Campbell, *Joker One: A Marine Platoon's Story of Courage, Leadership, and Brotherhood* (New York: Random House, 2009), 47 and 48.

5. For leading expressions of this critique, see *Network of Concerned Anthropologists: The Counter-Counterinsurgency Manual; or, Notes on Demilitarizing American Society* (Chicago: Prickly Paradigm Press, 2009), especially Marshall Sahlins's preface, i–vii; David H. Price, *Weaponizing Anthropology* (Petrolia, CA: CounterPunch Press, 2011).

6. Political theorist Michael Walzer speaks of "a recurrent incident in military history" in which a soldier, obligated by the most fundamental war convention—to take the life of an enemy soldier—decides not to pull the trigger after recognizing his counterpart's humanity. George Orwell, for instance, fighting for the Republicans in the Spanish Civil War, had a "full view" of a "man" who, after jumping out of a Nationalist trench, ran along the top of the parapet, "half-dressed and holding up his trousers with both hands as he ran. I refrained from shooting at him," Orwell wrote: "I had come here to shoot at 'Fascists'; but a man who is holding up his trousers is not a 'Fascist,' he is visibly a fellow-creature, similar to yourself, and you don't feel like shooting at him." *Just and Unjust Wars: A Moral Argument with Historical Illustrations* (New York: Basic Books, 1977), 138–40.

7. Dunahoe quoted in Richard S. Lowry, *Marines in the Garden of Eden: The True Story of Seven Bloody Days in Iraq* (New York: Penguin Books, 2006), 80.

8. Evan Wright, *Generation Kill: Devil Dogs, Iceman, Captain America and the New Face of American War* (New York: Putnam, 2004), 175–76.

9. Ibid., 83.

10. Bing West, *The March Up: Taking Baghdad with the United States Marines* (New York: Bantam Books, 2003), 67.

192 Notes to Pages 101–109

11. The U.S. Army, Marine Corps, *Counterinsurgency Field Manual* (Chicago: University of Chicago Press, 2007), 80, 239.
12. Ibid., 244.
13. U.S. Marine Corps, *Small Wars Manual* (1940; Honolulu: University Press of the Pacific, 2005), 4-3, p. 5; 4-15, p. 13; 6-78, p. 48; 1-10, p. 18.
14. Campbell, *Joker One*, 48.
15. "Culture of ferocity" comes from a *Slate* article about and interview with General Mattis. John Dickerson, "A Marine General at War," *Slate*, April 22, 2010, http://www.slate.com/articles/life/risk/2010/04/a_marine_general_at_war.html (accessed February 10, 2013).
16. Mattis quoted in Thomas Ricks, *Fiasco: The American Military Adventure in Iraq* (New York: Penguin Press, 2006), 313.
17. Ibid., 314.
18. Journalist Robert D. Kaplan quotes a Marine general as saying: "Write all about us, warts and all. We don't hide anything. We want the world to know exactly what we're like." Robert D. Kaplan, *Imperial Grunts: The American Military on the Ground* (New York: Random House, 2005), 266–67.
19. Mattis quoted in Ricks, *Fiasco*, 313.
20. *Counterinsurgency Field Manual*, 245.
21. Peter Baker, "Top Officers Fear Wide Civil Unrest," *Washington Post*, March 18, 2003.
22. General Conway spoke of "recocking" in "Briefing on the First Marine Expeditionary Force in Iraq," September 9, 2003. United States Department of Defense News Transcript. http://www.au.af.mil/au/awc/awcgate/dod/tr20030909-0658.htm (accessed February 26, 2013).
23. General Mattis, foreword to *Operational Culture for the Warfighter: Principles and Applications* (Quantico, VA: Marine Corps University Press, 2008), xi.
24. For an accomplished critical account of the Marines in Haiti, see Mary A. Renda, *Taking Haiti: Military Occupation and the Culture of U.S. Imperialism, 1915–1940* (Chapel Hill: University of North Carolina Press, 2001).
25. The point comes through in Nick Turse's book, *Kill Anything That Moves: The Real American War in Vietnam* (New York: Metropolitan Books, 2013).
26. See, for example, Smedley Butler, *War Is a Racket* (New York: Round Table Press, 1935).
27. General John F. Kelly, foreword to *U.S. Marines and Counterinsurgency in Iraq* (Quantico, VA: Marine Corps University Press, 2009), vii.
28. Bing West, *The Strongest Tribe: War, Politics, and the Endgame in Iraq* (New York: Random House, 2008), 151.
29. General Bargewell's 104-page report has not been released to the public. Although it was obtained by the *Washington Post* and summarized in Josh White, "Report on Haditha Condemns Marines," *Washington Post*, April 21, 2007.
30. Renda, *Taking Haiti*.
31. Not all occupations are alike in every way. With the Marines in early twentieth-century Haiti, as Renda explains it, a race-based paternalism over whom the Americans saw as "savage" and "child-like natives" enabled the high-levels of violence, including atrocities, against insurgents and noncombatants. Racial hatred of Iraqis certainly existed among Marines in the Al Anbar occupation, although the paternalistic discourse found in Haiti simply didn't take on such pronounced forms in Iraq.
32. Connors is quoted in West, *Strongest Tribe*, 151.
33. Raymond Williams, *The Long Revolution* (London: Chatto and Windus, 1961).
34. Dexter Filkins, *The Forever War* (New York: Vintage Books, 2008), 115.
35. Ibid., 123.
36. "Video Game Generation Takes Its Games to War Zone in Iraq," *Lubbock Avalanche-Journal*,

June 2, 2005, http://lubbockonline.com/stories/010205/war_010205102.shtml (accessed February 12, 2013). For a firsthand account by the winner of a *Halo 2* tournament at Camp Fallujah, see the blog, *Midnight in Iraq: A Marine's Blog*, July 20, 2006, http://midnight.hushedcasket.com (accessed August 1, 2006).

37. West, *Strongest Tribe*, 156.

38. *Mental Health Advisory Team (MHAT) IV, Operation Iraqi Freedom 05-07, Final Report*, November 17, 2006, http://www.armymedicine.army.mil/reports/mhat/mhat_iv/MHAT_IV_Report_17NOV06.pdf (accessed March 14, 2013).

39. Ibid.

40. The *Final Report*'s findings are clear in establishing a close relationship between combat and acute stress: "The level of combat is the main determinant of a soldier's or marine's mental health status"; "Multiple deployers reported higher acute stress than first-time deployers"; "Marines who experienced high combat were 4.6 times more likely to screen positive for acute stress."

41. West, *Strongest Tribe*, 148–58; Thomas E. Ricks, *The Gamble: General David Petraeus and the American Military Adventure in Iraq, 2006–2008* (New York: Penguin Books, 2009), 3–8.

42. "Selected Testimony from the Haditha Investigation," *New York Times*, December 14, 2011, http://www.nytimes.com/interactive/2011/12/15/world/middleeast/haditha-selected-documents.html?ref=middleeast&_r=0 (accessed March 17, 2013).

43. Mattis is quoted in West, *Strongest Tribe*, 153.

44. Jack E. Hoban and Joseph Shusko, "The Ethical Warrior and the Combat Mindset," *Marine Corps Gazette* (May 2012): 80–84.

45. Max Boot, *The Savage Wars of Peace: Small Wars and the Rise of American Power* (New York: Basic Books, 2002), 334.

46. Mattis quoted in S. D. Griffin, "Read a Book, Get Ahead," *Leatherneck* (December 2007): 42–43.

47. La Toya T. Graddy, "Reading Program Revamped," *Leatherneck* (April 2005): 44.

48. Donald R. Gardner, "Raising the Mental Bar," *Marine Corps Gazette* (April 2005): 13–14.

49. Bing West, *The Village* (1972; New York: Simon and Schuster, 2003), 13.

50. Ethan Hoalridge, "Gates of Fire: Marines Mirror Spartans," *Leatherneck.com*, November 16, 2007, http://www.leatherneck.com/forums/showthread.php?57181-Gates-of-Fire-Marines-mirror-Spartans (accessed May 9, 2013).

51. Steven Pressfield, *Gates of Fire* (New York: Bantam Books, 1998), 85.

52. Richard Lee, "Richard Lee's Spring 2000 Interview with Steven Pressfield," *Historical Novel Society*, http://historicalnovelsociety.org/gates-of-fire (accessed May 5, 2013). In another interview, Pressfield describes the book's explicit themes: "*Gates of Fire* has a theme, and the theme is courage. It's also very much about the camaraderie of fighting men and of the warrior ethos. Believe me, this is still alive and well, despite all P.C. efforts to exile it into the past. Today's Marines and soldiers, however, like the rest of us, are woefully undereducated. No one has studied the past, so we all feel as if we're the first people on the planet to be confronting the issues we're confronting. That's where a book like *Gates* fills a gap. Marines and Army guys read it and realize that the same stuff they're going through has been gone through by a lot of other warriors before them, and that those warriors and the societies they lived in had highly evolved codes of honor and conduct. It gives our young soldiers and Marines a longer historical perspective and inspires them that they're not alone and they're not the first; in fact, they're part of a long and honorable tradition of the profession of arms. It helps!" "Interview with Steven Pressfield," *goodreads*, June, 2008, http://www.goodreads.com/interviews/show/7.Steven_Pressfield (accessed May 10, 2013).

53. Pressfield, *Gates of Fire*, 128.

54. Paul L. Croom II, "A Lesson from Thermopylae," *Marine Corps Gazette* (January 2008): 54–56.

55. A vignette in the *Counterinsurgency Field Manual*—one that closely resembles the factual circumstances of the initial attack on Marines in Haditha—emphasizes how strong leadership can defuse a situation where marines might act outside the warrior ethos: the company commander "knew his Marines and understood the operational environment. He assessed the situation and acted aggressively to counter a dangerous situation that threatened mission accomplishment. By his actions, the commander demonstrated patience, presence, and courage." *Counterinsurgency Field Manual*, 243–44. The Professional Reading Program of the II Marines Headquarters Group assigned *Gates of Fire* and provided recommended study questions which included one that goes directly to "possession" and the responsibility of leadership to counteract it: "3. Compare how Dienekes kept his men from becoming 'possessed' after the battle against the Syrakusans and the Antirhonians with the following quote by General Dwight Eisenhower: 'There is a demon in every man. It is the duty of every officer to control that demon within himself, and in his men.'" "II MHG Professional Reading Program: *Gates of Fire*" (n.d.).

56. Pressfield, *Gates of Fire*, 128.

57. Croom, "Lessons from Thermopylae," 56.

58. Pressfield describes in a May 15, 2013, email correspondence with the author: "I got it into my head (along with my partner in Black Irish Books, Shawn Coyne) that I wanted to send something, a book, free to our troops overseas. So we put together 'The Warrior Ethos,' printed up 18,000 copies at our own expense and sent 'em overseas. That was an adventure in itself—the mailing. We approached the army, but they dropped the ball. Only the Marine Corps and the Army Special Forces responded right away—I'm talking about individual units in Iraq and Afghanistan, as well as the Big Marine Corps and Big SF—and said, Hell yeah, send the books.

59. Steven Pressfield, *The Warrior Ethos* (Los Angeles: Black Irish Entertainment, 2011), 34–35.

60. Ibid., 50, 48, 78, 90.

61. Kevin Powers, *The Yellow Birds* (New York: Little, Brown, 2012), 144.

62. Janine Mills, "Language and Culture in the Marine Corps," *Marine Corps Gazette* (August 2012): 66–69.

63. Marine Corps, *Vision and Strategy, 2025*, 2, http://www.onr.navy.mil/~/media/Files/About%20ONR/usmc_vision_strategy_2025_0809.ashx (accessed May 22, 2013).

64. *Counterinsurgency Field Manual*, 86.

65. Salmoni and Holmes-Eber, *Operational Culture for the Warfighter*, 18–19, 211.

66. Thomas E. Ricks, "Military Resilience, Suicide, and Post-Traumatic Stress: What's Behind It All?," *The Best Defense*, March 21, 2013, http://ricks.foreignpolicy.com/posts/2013/03/21/military_resilience_suicide_and_post_traumatic_stress_whats_behind_it_all (accessed May 24, 2013); Joseph Bobrow, "Learning from Marines about Military Suicides," *Huffington Post*, March 28, 2013, http://www.huffingtonpost.com/joseph-bobrow/learning-from-marines-abo_b_2972854.html (accessed May 24, 2013).

67. Staff Sergeant Danielle M. Bacon, "From Kinetics to Culture: 26th MEU Marines Learn Valuable Lessons," June 14, 2010, http://www.26thmeu.marines.mil/News/NewsArticleDisplay/tabid/2723/Article/65666/from-kinetics-to-culture-26th-meu-marines-learn-valuable-lessons.aspx (accessed May 28, 2013).

68. Consider, for instance, Marine statements in an article on the Infantry Immersion Trainer by Rick Rogers, "Marines Unveil Holographic Technology for Combat Training," *San Diego Union-Tribune*, January 15, 2008: "The Marine Corps is embracing breakthrough holo-

graphic technology to teach combat tactics and battlefield ethics at Camp Pendleton as troops there begin another major round of deployments to Iraq." Colonel Clarke Lethin, chief of staff for the I Marine Expeditionary Force, said that the system, as Rogers writes, "could help guide Marines through the tough process of making split-second battle decisions involving morality and legality. . . . 'As we go through the war, it's changing out there. There are more no-shoots than shoots,' Lethin said. 'We want to make sure that we are shooting the right people.' During the Iraq war, some incidents in the combat theater have embarrassed the Marine Corps and called into question its leadership and training on ethics." http://legacy.utsandiego.com/news/military/20080115-2142-bn15marines.html (accessed May 20, 2013).

69. Michael Gerson, "In Haiti, Empathy Is a Corps Competency," *Washington Post*, February 17, 2010, http://articles.washingtonpost.com/2010-02-17/opinions/36912356_1_american-marines-carrefour-2nd-marines (accessed May 19, 2013).

70. James Dao, "Ad Campaign for Marines Cites Chaos as a Job Perk," *New York Times*, March 9, 2012, http://www.nytimes.com/2012/03/10/us/marines-marketing-campaign-uses-chaos-as-a-selling-point.html?_r=0 (accessed May 2, 2013); Jason Del Rey, "With New Multimedia Campaign, Marines Shine Light on Gentler Side of Service," *Advertising Age*, March 9, 2012, http://adage.com/article/news/marines-shine-light-gentler-side-service/233223/ (accessed May 19, 2013).

71. http://www.youtube.com/watch?v=Wi8RTlFxcUI&list=PLRrmCq9BB1FBolfVjmbV-A2eikaGDGoQj&index=1.

72. http://www.youtube.com/watch?v=22b3LZIzEoE.

73. Dao, "Ad Campaign for Marines."

CHAPTER 5: **Posthuman Warfighting**

1. Robert Gates, George P. Schultz Lecture, San Francisco, August 12, 2010, United States Department of Defense, http://www.defense.gov/speeches/speech.aspx?speechid=1498 (accessed June 17, 2013).

2. Lieutenant Colonel Lloyd Freeman, "Can the Marines Survive?," *Foreign Policy*, March 26, 2013, http://www.foreignpolicy.com/articles/2013/03/26/can_the_marines_survive (accessed June 14, 2013). See also, Gregory A. Thiele, "The Marine Corps of the Future," *Marine Corps Gazette* (January 2011): 47–51.

3. In 2006, General Michael T. Moseley, U.S. Air Force Chief of Staff, said, "We've moved from using UAVs primarily in intelligence, surveillance, and reconnaissance roles before Operation Iraqi Freedom, to a true hunter-killer role." Mike Hanlon, "U.S. Air Force's First Hunter-Killer UAV Named Reaper," *Gizmag*, January 13, 2006, http://www.gizmag.com/go/6149/ (accessed July 2, 2013).

4. Michael Goldblatt in Joel Garreau, "Perfecting the Human," *Fortune* 151 (May 30, 2005), as quoted in Daniel McIntosh, "The Transhuman Security Dilemma," *Journal of Evolution and Technology* 21 (December 2010): 32–48.

5. Colonel James Laswell, quoted in P. W. Singer, *Wired for War: The Robotics Revolution and Conflict in the 21st Century* (New York: Penguin Press, 2009).

6. Ibid., 163.

7. Orson Scott Card, *Ender's Game* (New York: Starscape Books, 1985), 314, 322.

8. Ibid., 309.

9. Jack E. Hoban and Joseph Shusko, "The Ethical Warrior and the Combat Mindset," *Marine Corps Gazette* (May 2012), http://pqasb.pqarchiver.com/mca-members/doc/1013985247.html?FMT=FT&FMTS=ABS:FT:PAGE&type=current&date=May+2012&author=Hoban%2C+Jack+E%3B+Shusko%2C+Joseph&pub=Marine+Corps+Gazette&edition=&startpage=80-83&desc=The+Ethical+Warrior+and+the+Combat+Mindset (accessed April 16, 2013). Fur-

ther discussion of the Marine Corps Marshal Arts Program includes Captain Jamison Yi, "MCMAP and the Marine Warrior Ethos," *Military Review* (November–December 2004): 17–24.

10. Hoban and Shusko, "Ethical Warrior."

11. Warhammer website, http://www.games-workshop.com/gws/ (accessed April 20, 2013).

12. *Aliens: Colonial Marines* wiki, http://avp.wikia.com/wiki/Aliens:_Colonial_Marines_(2013_video_game/ (accessed April 20, 2013).

13. "Fragged," *Battlestar Galactica*, second season, originally aired July 29, 2005, Sci-Fi Channel.

14. James K. Sanborn, "Cameron: 'Avatar' Not Slam on Corps," *Marine Corps Times*, January 23, 2010, http://www.marinecorpstimes.com/article/20100123/NEWS/1230307/Cameron-8216-Avatar-not-slam-Corps (accessed April 23, 2013).

15. Adam Hadhazy, "The Science behind James Cameron's *Avatar*," *Popular Mechanics*, December 16, 2009, http://www.popularmechanics.com/technology/digital/fact-vs-fiction/4339866 (accessed April 23, 2013).

16. Robert A. Heinlein, *Starship Troopers* (1959; New York: Berkeley Publishing Group, 1987), 1. Biographical material on Heinlein from 337.

17. F. R. Thomas, "A Science Fiction Classic as a Training Manual," *Marine Corps Gazette* (September 1992): 78–79.

18. Ibid., 127.

19. Ibid., 1, 7.

20. Joe Pappalardo, "The Range Is Hot," *Popular Mechanics* (July–August 2013): 72–79.

21. David Axe, "Semper Fly: Marines in Space," *Popular Science*, December 18, 2006, http://www.popsci.com/military-aviation-space/article/2006-12/semper-fly-marines-space (accessed June 13, 2013); Sharon Weinberger, "The Very Real Plans to Put Marines in Space," *Popular Mechanics*, April 14, 2010, http://www.popularmechanics.com/technology/military/planes-uavs/plans-for-marines-in-space (accessed June 1, 2013).

22. Jeffrey L. Eby, "It Is Time for Exoskeleton," *Marine Corps Gazette* (September 2005): 33–35.

23. Major C. Travis Reese, "Exoskeleton Enhancements for Marines: Tactical-Level Technology for an Operational Consequence" (Master's thesis, United States Marine Corps School of Advanced Warfighting, 2010), 15.

24. A basis for this point is historian of science Donna Haraway's influential essay, "A Cyborg Manifesto: Science, Technology, and Socialist-Feminism in the Late Twentieth Century," in *Simians, Cyborgs and Women: The Reinvention of Nature* (New York: Routledge, 1991), 149–81.

25. John Keenan, "Editorial: Women in Combat," *Marine Corps Gazette* (March 2013): 3.

26. General Barrow's 1991 testimony before the Senate Armed Services Committee can be found in video form at http://www.mca-marines.org/leatherneck/gen-robert-h-barrow-women-combat.

27. B. L. Brewster and R. K. Wallace, "Let Us Fight for You," *Marine Corps Gazette* (June 2013), http://pqasb.pqarchiver.com/mca-members/doc/1372347839.html?FMT=FT&FMTS=ABS:FT:PAGE&type=current&date=Jun+2013&author=Brewster%2C+B+L%3B+Wallace%2C+R+K&pub=Marine+Corps+Gazette&edition=&startpage=67-70&desc=Let+Us+Fight+for+You (accessed July 3, 2013).

28. Ibid.

29. Mackubin T. Owens, "Women in Combat—Equal Opportunity of Military Effectiveness," *Marine Corps Gazette* (November 1992), http://pqasb.pqarchiver.com/mca-members/doc/206357055.html?FMT=FT&FMTS=ABS:FT:TG:PAGE&type=current&date=Nov+1992&author=Owens%2C+Mackubin+T&pub=Marine+Corps+Gazette+(pre-1994)&edition=&startpage

=32&desc=Women+in+Combat-Equal+Opportunity+or+Military+Effectiveness%3F (accessed July 3, 2013).

30. For instance, USMC master sergeant David K. Devaney argues that "we're not culturally ready," without recognizing the huge changes in sexual and gender politics that have occurred over the past several decades outside of the Marines. "Women in Combat Arm Units," *Marine Corps Gazette* (June 2012), http://pqasb.pqarchiver.com/mca-members/doc/1020697027.html?FMT=FT&FMTS=ABS:FT:PAGE&type=current&date=Jun+2012&author=Devaney%2C+David+K&pub=Marine+Corps+Gazette&edition=&startpage=62-64&desc=Women+in+Combat+Arms+Units (accessed July 3, 2013).

31. Steven Pressfield, *The Warrior Ethos* (New York: Black Irish Entertainment, 2011), 6.

32. DARPA website, "Tag Team Threat-Recognition Technology Incorporates Mind, Machine," http://www.darpa.mil/NewsEvents/Releases/2012/09/18.aspx (accessed July 5, 2013).

33. *2012 U.S. Marine Corps Science & Technology Strategic Plan*, https://www.movesinstitute.org/wp-content/uploads/2012/03/USMCSTStratPlan2012.pdf (accessed July 5, 2013).

34. Elisabeth Bumiller, "Air Force Drone Operators Report High Levels of Stress," *New York Times*, December 19, 2011, A8; Elisabeth Bumiller, "A Day Job Waiting for a Kill Shot a World Away," *New York Times*, July 30, 2012, A1, 12.

35. David Crane, "USMC Dragon Runner Mini Recon Robot (Ground Robot)/UGV for Urban Warfare Ops," *Defense Review*, January 14, 2005, http://www.defensereview.com/us-marine-corps-experimenting-with-new-recon-robot-for-urban-warfare-ops/ (accessed July 12, 2013).

36. D. M. Moreau, "Unmanned Ground Vehicles," *Marine Corps Gazette* (2004): 24, 26, 28.

37. Isaac D. Pacheco, "Maximum Impact: Battlebots Lead Impressive Field of New Gear," *Leatherneck* (September 2007): 44–50.

38. Arkin quoted in Robert S. Boyd, "Pentagon Exploring Robot Killers That Can Fire on Their Own," *McClatchy Newspapers*, November 24, 2009, http://www.mcclatchydc.com/2009/03/25/v-print/64779/pentagon-exploring-robot-killers.html.

39. The Gladiator Project, National Robotics Engineering Center, Carnegie Mellon University, http://www.rec.ri.cmu.edu/projects/TUGV/ (accessed July 12, 2013). See also James W. McCains, "The Marine Corps Robotics Revolution," *Marine Corps Gazette* (January 2004): 34–37.

40. Ibid., 34.

41. John Bohlin, "Robotics on the Battlefield," *Marine Corps Gazette* (September 2008): 31–36.

42. Pike quoted in Faye Flam, "Getting Robots of War to Act More Naturally," *Philadelphia Inquirer*, June 12, 2008, http://articles.philly.com/2008-06-12/news/25248986_1_new-robots-ants-swarms.

43. Ron Arkin, *Governing Lethal Behavior in Autonomous Robots* (2009), http://www.cc.gatech.edu/ai/robot-lab/online-publications/formalizationv35.pdf (accessed July 14, 2013).

44. Christopher Coker, *Warrior Geeks: How 21st Century Technology Is Changing the Way We Fight and Think about War* (New York: Columbia University Press, 2013), 177.

45. Paul W. Kahn, "The Paradox of Riskless Warfare," *Yale Law School Faculty Scholarship Series* (2002), Paper 326, http://digitalcommons.law.yale.edu/cgi/viewcontent.cgi?article=1325&context=fss_papers (accessed August 1, 2013). Political theorist Michael Walzer made the point about NATO attacks in 1999 on Serbian forces in Kosovo: "You can't kill unless you are prepared to die." NATO "cannot launch a campaign aimed to kill Serbian soldiers, and sure to kill others too, unless they are prepared to risk the lives of their own soldiers. They can try, they ought to try, to reduce those risks as much as they can. They cannot claim, [and] we cannot accept, that those lives are expendable, and these not." *Arguing about War* (New Haven: Yale University Press, 2004), 101.

46. Ibid.
47. Ibid.

CHAPTER 6: Synthetic Visions of War

1. Craig M. Cameron, *American Samurai: Myth, Imagination, and the Conduct of Battle In the First Marine Division, 1941–1951* (New York: Cambridge University Press, 1994), 23–24.
2. Ibid., 23.
3. Stephen Crane, *The Red Badge of Courage* (1895; repr., New York: Vintage Books, Library of America, 1990), 65; E. B. Sledge, *With the Old Breed: At Peleliu and Okinawa* (New York: Oxford University Press, 1981), 100.
4. John Keegan, *The Face of Battle* (New York: Viking Press, 1976), 128.
5. Andrew J. Bacevich, *The New American Militarism: How Americans Are Seduced by War* (New York: Oxford University Press, 2005), 6.
6. For a general introduction to biopolitics, see Michele Foucault, *Society Must Be Defended: Lectures at the College de France, 1975–1976* (New York: St. Martin's Press, 1997).
7. Arundhati Roy, "The Algebra of Infinite Justice," *Guardian*, September 29, 2001, http://www.theguardian.com/world/2001/sep/29/september11.afghanistan (accessed September 29, 2014).
8. For the all-volunteer force, Beth Bailey, *America's Army: Making the All-Volunteer Force* (Cambridge, MA: Harvard University Press, 2009).
9. Evan Wright, *Generation Kill: Devil Dogs, Iceman, Captain America and the New Face of American War* (New York: Putnam, 2004), 24.
10. Tom Engelhardt, "Drone Race to a Known Future," *TomDispatch.com*, November 10, 2009, http://www.tomdispatch.com/post/175155/tomgram:_droning_on (accessed September 29, 2014).
11. Thank you to Nicholas Schlosser, historian at the Marine Corps History Division, Quantico, for making this point and examples clear.
12. Bing West, *The March Up: Taking Baghdad with the United States Marines* (New York: Bantam Dell, 2003), 124.
13. Training and Simulation, Northrup Grumman, http://www.northropgrumman.com/Capabilities/TrainingandSimulation/Pages/default.aspx (accessed October 5, 2014). See also, Allen McDuffee, "A Holodeck Videogame Designed to Train Soldiers," *Wired*, January 28, 2014, http://www.wired.com/2014/01/holodeck/ (accessed October 5, 2014).
14. David Vergun, "'Live Synthetic' Army's Next Generation of Simulation," *www.army.mil*, March 19, 2014 (accessed October 9, 2014).
15. Joe Pappalardo, "The New Arsenal Weapons," *Popular Mechanics* (July–August 2014): 78–81.
16. "8 Civilians, including 3 Kids, Killed in US-Led Strikes on Syria," *RT*, September 23, 2014, http://rt.com/news/189872-children-killed-strike-syria/ (accessed October 12, 2014).
17. Mitchell Prothero and Jonathan S. Landay, "'Boots in the Air': U.S. Helicopters Return to Combat in Iraq for First Time," *McClatchy DC*, October 5, 2014, http://www.mcclatchydc.com/2014/10/05/242220/us-helicopters-pressed-into-combat.html (accessed October 11, 2014).
18. Nick Paumgarten, "We Are a Camera," *New Yorker*, September 22, 2014, 44.
19. Chad A. Stevens, *The Marlboro Marine* (New York: Mediastorm Productions, 2007).
20. Quote from Iraq War veteran's mother and suicide statistics from Robert Emmet Meagher, *Killing from the Inside Out: Moral Injury and Just War* (Eugene, OR: Cascade Books, 2014), 5, xii.
21. Ibid., xvii. Meager's *Killing from the Inside Out* includes discussion of Jonathan Shay's evolving concept of moral injury; the book also includes an afterword by Shay.

22. "The Marlboro Marine," *Mediastorm*, http://mediastorm.com/publication/the-marlboro-marine (accessed October 10, 2012).

23. In his memoir, *Packing Inferno: The Unmaking of a Marine* (Port Townshend, WA: Feral House, 2008), 81–82, Iraq War veteran Tyler Boudreau writes: "I think it's fair to say that the taking of human life is something we, as a species, are inherently reluctant to do. I've read that in books, but I've felt it in my gut too. But the soldier kills for a living—it is his reason for being. . . . But desensitization doesn't eliminate morality from the consciousness. It merely postpones cogitation. Sooner or later, when a man's had a chance to think things over, he will find himself standing in judgment before his own conscience."

24. Crane, *Red Badge of Courage*, 38.

25. Author interview of Colin Keefe, December 5, 2005.

26. Paul Fussell, *The Great War and Modern Memory* (New York: Oxford University Press, 1975), 22.

27. Anthony Swofford, *Jarhead: A Marine's Chronicle of the Gulf War* (New York: Scribner's, 2003), 6–7, 103, 70.

28. Kevin Powers, *The Yellow Birds* (New York: Little, Brown, 2012), 75, 43.

29. Phil Klay, "Redeployment," in *Redeployment* (New York: Penguin Press, 2014), 2.

30. Ibid., 15–16, 3, 8.

31. Ibid., 12–13.

32. Ibid., 13, 16.

Essay on Primary Sources

I based this book on primary sources of many kinds. As a work in contemporary history, the study draws from oral history projects simultaneous to the Iraq War, including my own, the Veterans Empathy Project. Now housed in a stand-alone interactive website, my interviews with U.S. veterans from the post–September 11 wars provided invaluable material on the Marine eye for battle, accounts that recreated points of view from atop a Humvee during the Iraq invasion, while on patrol in Fallujah during the long occupation, and when coming home to a nation reflexively proud of one's service yet immeasurably removed from and often indifferent to the actual content of that effort. The Veterans Empathy Project is meant to both bridge the yawning gap between civilian and military life and archive first-person accounts for further scholarship on the Iraq and Afghanistan wars.

I also benefited greatly from the U.S. Marine Corps' impressive oral history project on Operation Iraqi Freedom. Conducted by senior field grade officers trained in oral history methodology, these in-country interviews provided detailed accounts of the preparation, tactics, equipment, decision making, and effort that went into the spring 2003 Iraq invasion as well as documented the subjective experience, including the mind's eye, of marines who contributed to the operation. As few of the hundreds of interviews had been transcribed by the time of my research, I listened to the audiotaped recordings at the United States Marine Corps History Division in Quantico, Virginia, where hearing tone of voice and the dynamic between interviewer and subject only enriched my understanding of the human dimension of early twenty-first-century warfare. Interviews of pilots and other marines in the Third Air Wing proved unusually valuable in piecing together the combination of high technology and human expertise in the air war on Iraq.

The Iraq War has produced a veritable cascade of first-person written accounts of combat and service. From internet blogs to New York trade press titles, the hundreds upon hundreds of war memoirs published contemporaneous to the war itself know no analogue in American literary history. While my study would have been far thinner without these works, they do pose methodological challenges to documenting and finding meaning of the recent past. Like oral history, no one memoir can be taken as authoritative. One reads through them looking for patterns of experience, insight, and sense impression. With their limitations in mind, though, Iraq War memoirs (with the cascade ongoing) offer expansive detail and commentary on the social history—the lived experience—of U.S. war making.

Partially an institutional history, *Light It Up* draws from the myriad documents, publications, and self-definitional statements, appeals, and histories produced by the U.S. Marines; concerned with the ways the Corps shapes its recruits' worldview (understood in the colloquial sense), my study takes up the practices, stories, and assumptions embedded in this most idiosyncratic branch of the armed forces. This book is largely a work in cultural history. It focuses on texts paradigmatic to the Corps—recruitment commercials, for instance—as well as on underlying discourses that give coherence and direction to the Marines, and then analyzes those cultural excretions for the cognitive and emotional content and framework they bestow upon those who serve under the eagle, globe, and anchor banner.

On top of this standard approach to cultural history, I developed a research program that gets at the optics of the early twenty-first-century U.S. military and, more specifically, at the ways of seeing that the Marine Corps has cultivated for fulfilling its role as America's shock troops. My study turns on moving from Marine worldview in the general sense of mindset or outlook determined by an institutionally derived set of values, to worldview in the literal sense of how the Corps controls and augments its warfighters' sight lines for optimal use of its superior firepower. Examining Marine projection of force meant to me uncovering the predetermined systems and techniques of seeing that would then be shone on Iraq and other U.S. enemies. An important part of this exercise in the history of military visual culture involved studying the high-tech weapons systems that, using ever more advanced optical sensors, further distanced and effectively separated killer from killed. Matching the specs of night vision goggles and the thermal sight of the Abrams tank, for instance, with marines' accounts of fighting and killing with the aid of these optical technologies made up a major strand of my research.

Of equal importance were identifying and examining the particulars of "everyday" visual culture that aggrandize and *image* the act of face-to-face killing in war. Early twenty-first-century American popular culture is militarized through and through. I took this commonplace in contemporary cultural criticism as a starting point of my research into the different ways that force is projected on the country's enemies. With an elaborate and vivid visual popular culture that dramatizes warfare—and that privileges and finds pleasure in its violence—the United States has been able to heavily arm human subjects shaped by such content and place them overseas where they can "act out" the images that have already played before their eyes. Throughout most of the twentieth century, Hollywood combat films were an unusually effective means of making war through the mind's eye. And while they remain an active part of American force projection, new cultural forms have gained even greater prominence during this digital age. Therefore, a leading edge of my research has taken place on the internet and within the hugely influential subculture of video gaming. YouTube videos shot by marines in battle and on patrol and then edited and uploaded from camp make up some of the most unusual research for this book. YouTube war porn videos prompted thousands of viewer responses that document the patterns and layers of meaning people find in this new and disturbing genre of motion pictures. While replete with its own serious limitations as primary source material, this corpus of what viewers get from and look for in combat videos provides a rich answer to the cultural historian's long-standing dream of written record of cognitive and emotional responses to crucial texts.

Index

A-10 Warthog, 20
AC-130 Gunships, 68
acute stress, 111, 116, 119–20, 193n40. *See also* mental health; post-traumatic stress
Advanced Combat Optical Gunsight (ACOG), 31
Advanced Tactical Aerial Reconnaissance System (ATARS), 22
Afghanistan War, 150; all-volunteer force in, 153; and Cameron, 132; and combat video games, 88, 91, 92; and cultural awareness training, 119, 121; human forces in, 125, 127, 128, 155; and literature, 172, 174; marines urinating on corpses in, 49–50; and Matthis, 43; and noncombatant killings in, 116, 117; and Orth, 60; and propaganda, 40, 49; records of experiences in, 167; and Rumsfeld, 14; and soldiers as protectors, 140; suicides in, 169; and SUSTAIN, 135; and "Toward the Sounds of Chaos" campaign, 123; UAVs in, 16, 142, 143; and urban warfare training, 88; visual culture during, 159; and war porn, 164
air power, 145, 158; and attacks on ISIS, 164; and bombing as abstract idea, 13; and civilian casualties, 25, 37; and close air support, 21–22, 24, 29; and distancing of killer from killed, 13; and drones, 163; and empathy, 36; and expeditionary warfare, 134–35; and Fallujah, 68; and *First to Fight*, 92; and Iraq War, 1, 13, 15, 18–19, 20, 21, 25, 26. *See also* drones; unmanned aerial vehicles
Al Anbar Awakening, 103, 106
Al Anbar Province, 67, 68, 104; and counter-insurgency, 97–98; and cultural sensitivity, 99; Humvee patrols in, 109; infrastructure of, 104; and Mattis, 102; occupation of, 112; and U.S. Army, 98, 99. *See also* Fallujah; Haditha
Al Asad Base, 104
Al Aziziyah, 30, 34, 35
Al Budayr, 30
Alexander, Joseph H., 2, 106
Aliens (1986), 132
Aliens: Colonial Marines (video game), 132
Al Jazeera, 166
Allawi, Hussein, 108
all-volunteer force, 2, 153
Al Qaeda, 1, 142
Al Qaeda Iraq, 67, 104
American empire, 2, 51, 57, 106
America's Army (video game), 77
Amos, James F., 22, 122
amphibious assault landing, 127
Amphibious Assault Vehicles, 68, 134
An Nasiriyah, 12, 18, 24, 97
AN/PEQ-2 target laser (Infrared Target Pointer Illuminator/Aiming Light), 32
anthropology, 119, 127
anti-air defenses, 135
Apache attack helicopters, 164; and civilian casualties, 49–50, 52
Apocalypse Now (1979), 48, 60
arcade business, 73
Arkin, Ronald, 143, 146
Armored Caterpillar D9 Bulldozers, 68
artificial intelligence, 131, 141
Atari, 73
A-Team, The (1983–86; television series), 93
atomic bombs, 18
automated weaponry, 2, 148, 155, 160, 162, 163
AV-8B Harrier, 21, 22, 23, 125
Avatar (2009), 132–33

B-2 stealth bombers, 20
Bacevich, Andrew J., 153; *The New American Militarism*, 152
Baghdad, 26, 34, 88
Bailey, Beth, xi
Banana Wars, 21, 97
Bargewell, Eldon, 107, 112
Barrow, Robert H., 137–38
Barwanah, 104
Bassinger, Jeanine, 46
battlefield: visibility of, 15; visualization of, 5–7. See also war and combat
Battle Los Angeles (2011), 39, 131
Battlestar Galactica (2004–9; television series), 132
Battlezone (video game), 73–74
Baudrillard, Jean, 45
Bayonet Training (1938), 156
Beirut, 91, 93
Belgium, 38
Bell, David, 120
Belleau Wood, battle of, 149–51, 175
Berg, Nick, 53
biopolitics, 152–55
Black Hawk Down (2001), 79–80, 81, 158
Blackwater contractors, 65, 68–69
Blue Force Tracker, 15, 31, 157
"Bodies" (song), 54–55
body implants, 142
Bohlin, John, 145–46
bombing, as abstract idea, 13
Boot, Max, 113
boot camp, 4–5, 6, 139, 160. See also training
Born on the Fourth of July (1989), 47–48
Boudreau, Tyler, *Packing Inferno*, 199n23
Bourke, Joanna, *An Intimate History of Killing*, 44
Bowden, Mark, *Black Hawk Down*, 87, 88, 158
Boyd, Thomas, *Through the Wheat*, 113
Bradley Infantry Fighting Vehicle, 68, 73
Brewster, B. L., 138, 139
Breyer, Stephen, 70
Brown v. Entertainment Merchants Association, 70
Broyles, William, 48
Buhl, Willard A., 21, 31, 69
"Bullet with a Name on It" (song), 54

"Bullet with a Name on It" (video), 54, 56, 57, 64
Burns, Ed, 59
Bush, George W., 1, 18–19, 57, 162
Bushmaster 25 mm chain gun, 30
Butler, Smedley, 99, 106

Call of Duty (videogame), 63–64
camcorders, 165
cameras, helmet, 8, 40, 165
Cameron, Craig M., *American Samurai*, 43–44
Cameron, James, 132–33
Camp Abu Ghraib, 69
Camp Baharia, 69
Campbell, Donovan, 99, 102
Camp Fallujah, 69, 109
Camp Lejeune, North Carolina, 76
Camp Pendleton, California, 5, 39, 40, 76
Caputo, Philip, 38, 45; *A Rumor of War (1977)*, 113
Card, Orson Scott, *Ender's Game*, 71–73, 74, 75, 129–30, 134
Cariker, Tom, 112
Carnegie Mellon University, National Robotics Engineering Center, 143, 144
cell phones, 157
Chapman, James, 120
charge-coupled device television (CCD-TV) visual sensor, 22–23, 24
Chechen rebels, 49, 87
Chechnya, 88
Chessani, Jeff, 112
Chiarelli, Peter, 107
Chosin Reservoir, 67
city mock-ups, 5, 86, 87. See also simulations; training
civilians: and Al Anbar occupation, 98; armed fighters among, 67; and city mock-ups, 86; and counterinsurgency, 97; and dehumanization, 53; and drones, 164; empathy for, 11, 100–101; in Fallujah, 65, 68; and *First to Fight*, 91; in Haditha, 103, 105, 106–12; and insurgents, 104; killing of, 13, 19–20, 25, 37, 58, 65, 68, 105, 106–12, 164; needs of, 97; and precision guided munitions, 19–20; respect for, 113; and robot warriors, 146; and

shock and awe, 19; and *Six Days in Fallujah*, 94; training for operations amid, 5; training in peaceful interactions with, 120; as victims, 49–50, 52. *See also* Iraqi people
"Climb, The" (2001; Marine Corps commercial), 122
clinical distress, 142. *See also* acute stress
Close Combat: First to Fight (video game), 65–66, 77–80, 81, 82, 87–88; authenticity of, 92, 93; and Hollywood, 79–80; and Marine Corps, 66, 88, 89–94; and second battle of Fallujah, 66; and urban combat, 78, 80, 89–94
Close Combat: Modern Warfare (video game), 92
CNN, 17
Cobra helicopters, 21, 51, 125
cognition, 15, 128; bodily, 86, 156, 160–62, 166
Cognitive Technology Threat Warning System, 141
cognitivism, 82–83
Coker, Christopher, 147
Cold War, 47, 73, 75, 87, 131
colonial force, 2, 97
Columbine killings, 77
combat. *See* war and combat
combat stress, 120. *See also* acute stress; post-traumatic stress
"Combat Town" (Twentynine Palms, California), 5, 86
Combat Training Centers, 87
combat trauma, 169. *See also* acute stress; post-traumatic stress
Combined Anti-Armor Attack Team (CAAT), 27, 33, 34, 35
Common Imaging Ground/Surface Stations, 22
Connors, Timothy, 38, 108
Conway, James T., 21, 32, 105, 118
Coppola, Francis Ford, 48
counterinsurgency, 11, 25, 67, 87, 93–94, 99, 102, 106; and Combined Action Platoon, 114–15; and cultural awareness, 120; and culture, 118–19, 120; and empathy, 100, 101; and face recognition optics, 142; and humanitarianism, 98, 99, 101, 103; illegitimacy of, 106; and intelligence, 101; and Iraq War, 10, 96, 97; and Mattis, 99, 102–3; and turning off of killing switch, 96; and Vietnam War, 114

Crane, Stephen, *The Red Badge of Courage*, 151, 165, 170, 172, 174
Croom, Paul L., II, 116–17
Crowther, Bosley, 46, 47
cruise missiles, 18, 20
culture, 120, 177n4, 201–2; awareness of, 98, 106, 109, 127; and battle conduct, 64; and battlespace, 119; causal loop between war and, 11; and counterinsurgency, 118–19, 120; digital, 157, 158, 166; and eye for battle, 7; and gender roles, 139; and Haitian earthquake, 121–22; Marine, 99, 177n4; militarization of American, 58; popular visual, 7–8; sensitivity to, 99, 127; training in, 118–19; and violence, 1; visual context of, 41; and war, 8
cybernetics, 128, 136, 137, 141, 159
cyborg, 142, 155, 163

Daly, Dan, 149
Danelo, David J., *Blood Stripes*, 9
DARPA. *See* Defense Advanced Research Projects Agency (DARPA)
Davis, Richard Harding, 190n69
"D-Day Fallujah" (television documentary), 61–63
Deadliest Warrior (2009–12; television program), 58
death, 36, 158, 169, 171; and Marine Corps Museum, 150, 151, 153; risk of, 155; in videos, 53, 55. *See also* killing; suicide; violence
Defense Advanced Research Projects Agency (DARPA), 128, 141
defense industry, 158; and video games, 75–76, 82
Defense Video and Imagery Distribution System, 39
Department of Defense, 49, 73, 75, 87; cable channels of, 58; and exoskeleton, 136; and network-centric warfare, 14; and simulations, 70–71, 76; and SUSTAIN, 135; and women in combat, 137
Destineer Corporation, 65–66, 77, 79, 87–88, 89–94
"Die Mother Fucker Die" (song), 53

Index

digital cameras, 8, 40, 50, 158, 165, 185n37
digital communications, 14
digital culture, 157, 158, 166
digital fire control computer, 29
digital information technology, 128
digital media, 40
digital technology, 76, 82, 156–57
Discovery Channel, 58
Dispatches (1977), 46
distributed operations, 135, 141
Diwaniyah, 12
Dogfights (2006–7; television program), 61
Dolan, Jennifer, 23
Doom (video game), 69, 77
Dragon Eye, 17, 31, 156, 157, 181n53
Dragon Runner, 143, 144
drones, 128, 131, 142, 162; advent of, 11; and attacks on ISIS, 164; and civilian deaths, 164; increasing utilization of, 155. *See also* unmanned aerial vehicles
Drowning Pool, 54
dual-life theory, 148
Dunahoe, Brent, 100
Dwan, Allan, 47

E-8C aircraft, 16
Eby, Jeffrey L., 135–36
Eckhart, Aaron, 39
empathy, 36, 106, 118, 201; for civilians, 11, 100–101; and counterinsurgency, 100, 101; at distance, 36; and *First to Fight*, 94; and robots, 146; tactical, 101; and war, 129
"Enduring Freedom: The Opening Chapter" (2003; promotional film), 40
enemy: dehumanization of, 36, 52–53, 100, 130; denigration of, 28; fetishization of, 53; human connection with, 170; humanity of, 10–11, 191n6; mediated perception of, 131; morale and discipline of, 93; and moral guilt, 147–48; perception and understanding of, 29, 129; respect for, 113, 118; as unknowable other, 53; unknown, 55; as worthy, 69. *See also* Iraqi insurgents; Iraqi military
Engelhardt, Tom, xi, 155
entertainment industry, 39, 70, 73, 76, 158. *See also* films; video games

Ergo, Mike, 94
Ermey, R. Lee, 61
Espera, Antonio, 100–101
ethics, 58, 60, 103, 110, 112–13; and battle robots, 145; and killing, 129, 130, 147, 159; and moral injury, 169, 175; and Obama, 163; and policing, 147–48; and Pressfield, 117; and robots, 146–47, 148; and self-possession, 159; and subjectivity, 167; and training, 4, 130; and USMC, 112–13
exoskeleton, 128, 133, 135–37, 141, 163
Expeditionary Fighting Vehicle, 134
eye, 22, 160; as automated, 159; as Cartesian tube, 8; and technology, 7; and training, 8, 9; violence located in, 3, 141. *See also* seeing/sight

F-15C fighter, 20
F-16 fighter, 20
F-35B Lightning II fighter, 134–35
F-117 fighter, 20
F/A-18 Hornet, 20, 21, 22, 125, 135
Facebook, 164
face recognition optics, 142
facial expression, 41–42, 55, 80, 81, 90, 189n52
Fair, Eric, 170
FalconView, 6–7, 157
Fallujah, 65–69, 106, 108; Blackwater killings in, 68–69; and Gladiator, 144; Jolan district of, 62
Fallujah, first battle of, 65, 103, 104
Fallujah, second battle of, 9, 55, 96, 103, 104; course of, 67–69; and "D-Day Fallujah," 61–63; and film staging, 38–39; and *First to Fight*, 66; and Haditha killings, 111, 112; mock-ups used in preparation for, 5; shooting of insurgent in, 65; video about, 50–51. *See also* Al Anbar Province
Fedayeen fighters, 28, 29
feminism, 154–55
Fick, Nathaniel, 59; *One Bullet Away*, 43
Fightin' Marines (comic books), 43
Filkins, Dexter, *The Forever War*, 108–9
films: about Belleau Wood, 149–51; and bodily movement, 38; context construction by, 44; as creating eye for battle, 38; and desensitiza-

tion to violence, 44; determination of war making by, 41–42; facial expressions in, 41–42, 80, 81; as idealizing war, 156; influence of, 156; about Iraq War, 48–49; killing in, 43–44; and nonparticipatory viewing, 84; point-of-view shot in, 38, 81, 149–50; and preconsciousness, 38; as shaping perceptions, 38–39; soundtracks of, 80; training through, 156; about Vietnam War, 57; visual culture of, 155; and warriors, 1; about World War II, 43, 46–47, 57. *See also* Hollywood films
"First, do no harm" slogan, 102
First Amendment, 70
First Annual "Ben Hur" Memorial Chariot Race, 69
First to Fight (slogan), 8–9
Flatworld, 9, 76, 160–61
Flip cameras, 165
Foley, James, 166, 185–86n52
force projection, 17, 32, 155, 158; as term, 7
"For God and Country: A Marine Sniper's Story" (2006; television program), 60–61
Fort Polk, Louisiana, 87
forward air controllers, 22, 24, 68
forward armed reconnaissance, 21
forward-looking infrared (FLIR), 22, 68
four-man fire team tactics, 65, 77, 90, 158
Fox Television, 58
Franks, Tommy, 1, 13, 18–19, 20
fratricide, 15, 24–25, 31
freedom of speech, 70
Freeman, Lloyd, "Can the Marines Survive?," 127
Full Metal Jacket (1987), 49, 51, 61, 69–70, 80, 81
Full Metal Jousting (2012; television program), 61
Full Spectrum Warrior (video game), 82
Fussell, Paul, *The Great War and Modern Memory*, 171

Gardner, Donald R., 114
Gates, Robert M., 127
GBU-16 (Guided Bomb Unit), 23
gender, 136–37, 142. *See also* women

Generation Kill (2008; television program), 41–42, 58, 59–60
George Air Force Base, Victorville, California, 5
Gerson, Michael, 121–22
G.I. Joe action figures, 43
Gladiator, 143–45
Global Hawk, 15–16
global positioning system (GPS), 6, 15, 31. *See also* satellites
Global War on Terror, 108, 123
Gomer Pyle U.S.M.C. (1964–69; television series), 40
Google Glass, 8
GoPro camera, 165
Gorgon Stare, 7
Grabowski, Thomas, 16
Grapes, Jesse, 63
Gray, Alfred M., 114
Green Zone (2010), 48
Grossman, David, 28, 100; *On Killing*, 3, 4, 44
Grozny, 87
Guadalcanal Diary (1943), 45, 46
Gulf War (1990–91), 15, 55, 156
Gung Ho (1943), 46–47
gun-sight footage, 55–56

Haditha, 103, 104, 105, 106–12. *See also* Al Anbar Province
Hagee, Michael, 114
Haiti, 106, 107–8, 119, 121–22
Haitian Operation Unified Response, 122
Halo 2 (video game), 85, 109
Halo 3 (video game), 92
Halter, Ed, 77
Hamlin, Denny, 82
Hand to Hand Combat in Three Parts (1942), 156
Hanson, Michael, 32
Harris, Eric, 77
Hasford, Gustav, *The Short-Timers*, 113
HBO, 41, 42, 59, 64
Head Up Display, 141
heavy metal, 50, 53
Heinlein, Robert A., *Starship Troopers*, 133–34, 135, 136, 142
Hemingway, Ernest, 37
high technology, 14, 29

high-technology mapping, 6–7
high-technology sensors, 155, 158
high-technology weapons systems, 127, 128, 153, 158, 202
high-tech optics, 29, 137
Hiroshima and Nagasaki, 18
History Channel, 61, 62, 94
hivemind, 141
Hoalridge, Ethan, 115
Hoban, Jack E., 130
Hollywood films, 1, 70, 75, 76, 156, 160, 202; and *First to Fight,* 79–80; influence of, 45–49; influence of, on video games, 79–82; and Institute for Creative Technologies, 76; and Marine Corps, 39–40, 44, 45, 46, 47, 49; as shaping perceptions, 38, 45–46; and war porn, 57. *See also* films
Holmes-Eber, Paula, *Operational Culture for the Warfighter,* 119
Holodeck, 85, 161
Holt, Lester, 60
Homeland (2011–; television program), 59
Home of the Brave (2006), 48
Homer, *The Iliad,* 37, 69–70
honor, 89, 112, 113, 115, 117, 118, 120, 128
Huck, Richard, 112
Hue City, Vietnam, 63, 67
Huizinga, Johann, *Homo Ludens,* 66–67
humanitarianism, 97, 130, 131, 148; and counterinsurgency, 98, 99, 101, 103; and empathy, 101; and Haiti, 119, 121–22; and Mattis, 103; and recruitment commercials, 119, 122, 123, 124, 125; and robot warriors, 147
human rights groups, 164
Hurt Locker (2008), 48
Hussein, Saddam, 1, 17, 18, 19, 105
hypermasculinity, 2, 7, 44, 58–59, 140, 141

imaging, 22, 144; and high-tech systems, 6–7; and radar satellites, 15; of violence, 8; and visualization, 5–7
improvised explosive devices (IEDs), 97, 105, 106, 129, 143
Inchon, 67
Infantry Immersion Trainers, 85

Infrared Target Pointer Illuminator/Aiming Light, 32
Institute for Creative Technologies. *See* University of Southern California, Institute for Creative Technologies
intelligence, surveillance, and reconnaissance (ISR), 23, 24, 128
Interservice/Industry Training, Simulation, and Education Conference (I/ITSEC), 83
In the Valley of Elah (2007), 48
Iraq: air defenses of, 16, 18; ISIS in, 164; and oil, 19, 57–58; sectarian strife in, 26–27, 105; stress of occupation of, 110–12; and weapons of mass destruction, 51, 57
"Iraq: Bird Is the Word" (video), 49, 55, 56
Iraq Body Count study, 20
"Iraqi Freedom—Chapter II" (2003; promotional film), 40
Iraqi insurgents, 40, 106; and "D-Day Fallujah" television program, 62; and high-tech warfare, 14; and noncombatants, 104, 110; and war porn, 110; will of, to fight, 93. *See also* enemy; Fallujah
Iraqi military, 15, 20, 22; marksmanship of, 28–29, 32, 35; will of, to fight, 29
Iraqi people, 1, 24, 30, 104; devaluation of, 107, 108; freeing of, 57; hostility toward, 110; as human beings, 10–11; humanitarian aid to, 49; interaction with, 109–12; memories of killing of, 27; objectification of, 53; and Orth, 60; peaceful relations with, 104; and shock and awe, 19; terrorizing of, 25–26; treated as insurgents, 110. *See also* civilians
Iraqi Republican Guard, 15, 16, 22, 26, 28, 34
Iraq War, 123, 125, 140, 143, 152; air campaign in, 13, 18–20; American invasion in, 6, 14; counterinsurgency stage of, 10, 96, 97; course of, 1–2; and digital culture, 158; and digital media, 40; and empire, 106; entertainment value of, 41; films about, 48–49; human forces in, 125, 127, 128, 155; and literature, 37–38, 172, 174; and Marine combat camera teams, 39; motion pictures of, 165; network news coverage of, 57; and Obama, 164; opposition to, 1, 2; and primetime dramatic series, 57; records of experiences in, 167;

suicides in, 169; on television, 57–64; training for, 5; and urban warfare training, 88; value of, 26–27; videos about, 42, 49; visual culture during, 159; and war porn, 53, 164; winding down of, 145
"Iraq War—Let the Bodies Hit the Floor" (video), 54–55
ISIS (Islamic State), 1, 53, 164, 166, 185–86n52
Islam, 119
Islamic extremists, 49, 93

James, William, 37
Japanese, 52
JDAMs (joint direct air munitions), 20
Johnson, Stephen, 112
Joint Surveillance Target Attack Radar System (STARS), 15, 16
J. Walter Thompson Agency, 122, 125

Kahn, Paul W., 147
Keefe, Colin, viii–ix, 167, 171; memories of, 12–13, 26–29, 32–36
Keegan, John, *The Face of Battle,* 151
Keenan, John, 137
Kelly, John F., 106
Khe Sanh, Vietnam, 67
kill boxes, 16, 21, 22, 23
killing, 127, 175; and Al Anbar occupation, 98; automated, 3; and autonomous robots, 147; and Belleau Wood film, 151; of civilians, 13, 19–20, 25, 37, 58, 65, 68, 105, 106–12, 164; and combat films, 43–44; and counterinsurgency, 97; and devaluing of non-American lives, 153; digitized models of, 159; and distancing of killer from killed, 3, 4, 11, 13, 23, 29, 30, 37, 108, 137, 141, 153, 155; and empathy, 100; and enemy's humanity, 191n6; and ethics, 129, 130, 147, 159; and eye, 3; and *First to Fight,* 91, 93; and gender, 138, 139; guilt over, 6, 36, 41, 60, 61, 167, 169; in Haditha, 103, 105, 106–12, 107; hesitation over, 3–4, 37, 146; and honor, 118; and human connection, 170; human control over, 146; and indifference to human cost, 13; instinct against, 28; and Keefe, 35; knowingness about, 154; and Lowry, 98;

motivation for, 45; nightmares about, 169–70; and Orth, 60; pleasure in, 46; proclivity for, 153; to protect life, 148; by remote control, 147; risk from, 145; and robots, 146; and seeing, 8; training for, 3, 4–5, 6; as unnatural, 139; in videos, 51, 52; and virtual presence, 76; and war porn, 41, 42, 160; and WWII combat films, 46–47. *See also* violence
killing switch, 159; as instilled, 9–10; turning off and on of, 105, 113, 153; turning off of, 10, 96, 97, 100, 104
Klay, Phil, *Redeployment,* 172–75
Klebold, Dylan, 77
Konami Corporation, 66
Kovic, Ron, *Born on the Fourth of July,* 47–48, 113
Krulak, Charles C., 77, 96, 97; "The Strategic Corporal," 88–89, 97; and three-block war, 103, 120
Kubrick, Stanley, 48, 81

Lafontant, Roosevelt, 135
laser-guided munitions, 14
lasers, 21, 22
Light Armored Vehicle, 30
"light it up," as expression, 7, 35
LITENING Pod, 22–24, 156
live-fire exercises, 86, 156
Live-Leak, 164–65
Lost Evidence (2004; television program), 61
Lowry, Tom, 98, 99, 167–68
ludology, 69, 70, 94, 101. *See also* video games; war and combat

M1 Abrams tank, 7, 29–30, 51, 68, 156, 157
M9 Armored Combat Earthmovers, 68
M203 40 mm grenade launcher, 32
M249 Squad Automatic Weapon, 32
Macedonia, Michael, 71, 83, 85
Mackin, Elton, *Suddenly We Didn't Want to Die,* 113
Mail Call (2002–9; television program), 61
Making Marines (2002), 58–59
Malay, Patrick J., 96, 98
Manaker, Ellen, et al., "Harnessing Experiential Learning Theory," 83

Manchester, William, *Goodbye, Darkness*, 113
MANswers ((2007–11; television program), 58
Marine, The (2006), 48
Marine Doom (video game), 77
Marine Light Armored Vehicles, 157
"Marines Get Some: Die MF Die" (video), 53–54
MarineTroopSupport (video uploader), 53
marksmanship, 4–5, 28–29, 32, 35, 87
Marshall, S. L. A., 28; *Men against Fire*, 3, 4
Martinez, Marco, 9–10
Massachusetts Institute of Technology, 73
Mattis, James, 5, 21, 43, 98–99, 102–3, 110, 145; and Marine literary culture, 113–14; on occupations, 105–6; rhetorical gifts of, 104
McCoy, B. P., 6
McMains, James, 144
Meagher, Robert Emmet, 169
memory: muscle, 82, 86, 156; procedural, 82, 85, 86
mental health, 109–12, 116. *See also* acute stress; post-traumatic stress; suicide
Mental Health Advisory Team for Operation Iraqi Freedom, 109–10
Messenger, The (2009), 48
Midway Sea Raider (arcade game), 73
Military Channel, 58
Military History Channel, 58
Miller, James Blake, 65, 168–70, 171
MK-19 grenade launcher, 34, 35
mock-ups. *See* city mock-ups
Mogadishu, 80, 87, 88, 158
moral injury, 169, 175. *See also* ethics
Moreau, Dave, 143
Morgan, Robin, 42
Moseley, T. Michael, 18
motion picture, 164, 165, 166. *See also* films
MSNBC, 60–61
MTV, 82
Multi-National Force-Iraq (MNFIRAQ), 49
Musselman, Mike, 30
MV-22 Osprey, 135

National Academy of Science (NAS), National Research Council, *Modeling and Simulation*, 75–76, 77, 84, 86–87
National Defense University, 17

National Imagery and Mapping Agency, 7
NBC, 60–61, 65
network-centric warfare, 7, 13–15, 18, 158, 178–79n4
Nicaragua, 106
night vision contact lenses, 163
night vision goggles, 7, 29, 31–32, 156, 157
Nintendo, 55
"No better friend, no worse enemy" slogan, 99, 102, 110
nonfiring rate, 3, 4, 28, 100, 146
Nonpoint, 54
Northrup Grumman, 161

Oakland, California, marine assault landing in, 158
Obama, Barack, 162–63, 164
O'Connell, Aaron B., xi, 4
O'Neil, Robert, ix–x
Operation Desert Storm, 16
Operation Enduring Freedom, 16
Operation Iraqi Freedom, 1, 2, 6, 14, 16, 33, 159; air power in, 18–19, 20–25, 29; civilian casualties in, 20; and digital culture, 157–58; and GPS, 15; ground war in, 20–22, 24–25; Opening Gambit, 21–22; USMC oral history project on, 201
Operation Southern Watch, 22
Operation Urban Warrior, 88, 91, 158
optical cameras, 68
optical systems, 22
optical technologies, 31–32
optics: advanced, 141; and Gladiator, 144; and hypermasculinity, 2; robotic, 146; and see-shoot training, 4; and solid objects, 141–42
optics of combat, 28, 65, 151, 152; as separating and mediating, 157; as synthetic, 155; and violence, 3; and visualization, 5
Orth, Matthew, 60–61
Orwell, George, 191n6
Osterman, Joseph L., 125
"Out of My Way" (song), 51
over-the-next-hill surveillance, 17
Over There (2005), 57–58
Owens, Mackubin T., 138–39

PackBots, 143
Pac Man (arcade game), 73
Pakistan, 142
Pape, Robert, 20
Patai, Raphael, *The Arab Mind*, 98
Patton, George, 104
Paumgarten, Nick, 165
Peebles, Stacey, 41
Pentagon Channel, 68
Pershing, John, 4
Persian Gulf War, 14
Petraeus, David H., 93, 99
pharmacology, 128
Pike, John, 146
play. *See* ludology; video games; war and combat
point-of-view shots, 38, 63, 81, 123, 149–50
Poletis, Nick, 97
policing, international, 147
Pong (video game), 73
population-centricity, 99, 107
pornography, 40, 63, 183n17; definitions of, 42. *See also* war pornography
posthuman warfighting, 11, 125, 128, 139–40, 145, 155, 163
post–September 11, 2001, era, 48, 58, 65, 156. *See also* September 11, 2001, attacks
post-traumatic stress, 82, 111, 119, 125, 130, 142, 175. *See also* acute stress; mental health
Powers, Kevin, *The Yellow Birds*, 172
precision-guided munitions, 15, 19, 20, 21, 22
precision munitions, 25
Predator drones, 7, 37, 52, 95, 164
Pressfield, Stephen, 140; *Gates of Fire*, 115–16, 117, 194n55; *Warrior Ethos*, 117–18
"Pride of the Nation" (2005; Marine Corps commercial), 40, 122–23
propaganda, 40, 41, 46
protection, of life, 140, 148

Quantico, 127, 156

race, 108, 119, 140, 154, 192n31
radical Islamists, 49
Ramadi, 104
rape, 42. *See also* sexual violence

rapid dominance, 17, 19
Ready-Team-Fire-Assist maneuvers, 78, 90, 92
Reagan, Ronald, 74–75
Real World (1992; television program), 82
Reaper drones, 164
reconnaissance, 6, 15–16, 17
recruitment, 39–40, 119, 122, 123, 124, 125; and biopolitics, 154–55; and video games, 160
reddit, 165
Reese, C. Travis, 136
remote weaponry, 1, 148, 155, 162, 163
Renda, Mary A., 107–8
Retreat Hell! (1952), 39
Rhodes, John, 87, 88
Ricks, Thomas E., 120; *Making the Corps*, 177n4
robots, 125, 128, 131, 140, 163; advent of, 11; combat, 164; optics for, 155; remote-controlled, 137
robot warriors, 143–47, 148, 155
Roddenberry, Gene, 131
Roosevelt, Teddy, 104
Rotberg, Ed, 73–74
Roy, Arundhati, "Algebra of Infinite Justice," 152–53
Rumsfeld, Donald, 14, 19, 178n4
Rumsfeld Doctrine, 158
Russell, Diana E. H., 183n17

Salmoni, Barak A., *Operational Culture for the Warfighter*, 119
Sands of Iwo Jima (1949), 39, 46, 47, 81
satellite mapping, 6, 7, 29
satellites, 14, 15, 31. *See also* global positioning system
Saving Private Ryan (1998), 79, 81
science fiction, 39, 70–71, 128–30
Scott, Ridley, 79, 158
Second Life (simulated reality site), 82
seeing/sight, 14–15, 158; and killing, 8; as mediated, 155; as military targeting, 8; technological extensions of, 155–56. *See also* eye
self-possession, 113, 117, 159, 167, 174
September 11, 2001, attacks, 88, 153, 163, 166. *See also* post–September 11, 2001, era

sexual equality, 127
sexuality, 40, 139
sexual violence, 42, 127, 140
Shakespeare, William, 37
Shay, Jonathan, 169
Sherry, Michael, xi, 13
Shia Muslims, 26–27
shock and awe, 1, 17–20, 33; as counterproductive, 19; and enemy's perceptions, 15, 17–18; failure of, 14; and Iraqi marksmanship, 29
Shootout (2005–6; television program), 61–64
Shootout: D-Day Fallujah (television program), 94
Showtime, 59
Shupp, Michael, 67
Shusko, Joseph, 130
Simon, David, 59
simulations, 65–66, 163; and behaviors in response to images, 84; and bodily cognition, 160; and Card, 129; and cognitivism, 82–83; and cultural awareness, 120; development of, 73–77; immersive, 160–62; and Institute for Creative Technologies, 76; interactivity of, 85; and Marine Corps, 70; and real world, 70–71; and storytelling, 76; of urban combat, 6, 158; and virtual presence, 75–76. *See also* training
Singer, P. W., 40
situational awareness, 15, 16, 31
Six Days in Fallujah (video game), 65, 66, 94–95
Sledge, E. B., 174; *With the Old Breed*, 113, 151
smart bombs, 19
smartglasses, 141, 142
smart phones, 40, 165
smart-scope rifle, 163
Smith, Ray L., 6
Sobchak, Vivian, *Carnal Thoughts,* 182n4
social inequalities, 154
Somalia, 87, 88, 100
Sontag, Susan, 182n4
Sotloff, Steven, 166, 185–86n52
Southern Focus, 18
Space Invaders (arcade game), 73
Space Invaders (video game), 73
Space Marines, 135
Spartans, 115–16, 117

Spike TV, 58
sports, 6, 67, 86, 156, 190n69. *See also* war and combat: as competition
Stahl, Roger, 57
Star Trek (1966–69; television series), 131
Star Trek: The Next Generation (1987–1994; television series), 76
Star Wars (1977), 79
Stevens, Chad A., 168
Stokes, Andrew Sean, 38–39
Stone, Oliver, 47–48
subjectivity: and combat videos, 45, 56, 57, 62; and counterinsurgency, 97, 101, 108; and films, 84; and *Full Metal Jacket,* 81; and simulations, 75–76; of warriors, 159–60, 167, 170, 171, 173–74, 176
suicide, 119, 120, 125, 169. *See also* mental health
suicide bombers, 98
Sulla, 102
Sunni Muslims, 26–27, 104, 106
SUSTAIN (Small Unit Space Transport and Insertion), 135
Sutton, Scott, 85
Swales, Sinque, 85
Swofford, Anthony, 174; *Jarhead,* 48, 113, 171–72
synthetic optics, 128
Syria, 1, 26, 164

Takacs, Stacy, 58
Taliban, 142
Tamte, Peter, 79, 80, 89, 90, 91, 92, 93, 94
Task Force Tarawa, 24
Taylor, Justin J., 84–85
Tell It to the Marines (1926), 39
Tendon-Assisted Rigid Exoskeleton, 163. *See also* exoskeleton
Terrain 2025, 161
Terrazas, Miguel, 107, 111
terrorists, 52, 163, 166
Terry, Jennifer, 45
TheCarl1001 (video commentator), 55
thermal sights, 7, 29, 30, 156, 157
"3rd Battalion, 1st Marines in Fallujah Iraq 2004" (2006; video), 50–51

Tikrit, 26
Till, Karen, 86
Tillman, Pat, 58
Tolstoy, Leo, 37
"Toward the Sound of Chaos" (USMC advertising campaign), 119, 122–24, 125
tracking scopes, 163
training, 101, 160; and cultural awareness, 120; and films, 156; for killing, 3, 4–5, 6; and live-fire exercises, 5; and marksmanship, 4–5; rehearsal and visualization in, 5; see-shoot, 3, 4, 146; and video games, 160; and videos taken by marines, 8. *See also* boot camp; city mock-ups; simulations; urban combat
Twentynine Palms, 98

Ullman, Harlan K., and James P. Wade, "Shock and Awe," 17–18, 19
United Kingdom, 14
University of Southern California, Institute for Creative Technologies, 9, 76, 82, 160
unmanned aerial vehicles (UAVs), 11, 15–16, 17, 31, 68, 125, 128, 142, 157. *See also* Dragon Eye; drones; Predator drones; Reaper drones
unmanned ground vehicles, 128, 143–45
urban combat, 5, 51; asymmetric, 87; and city mock-ups, 5, 86, 87; and combat simulations, 66; and "D-Day Fallujah" television program, 63; and Fallujah, 68–69; and *First to Fight*, 78, 80, 89–94; and four-marine squad, 88; and Gladiator, 144; likelihood of, 88; and low-level decision making, 88–89; simulations of, 6, 158; and video games, 64, 65, 77, 82, 87–88, 158. *See also* war and combat
Uris, Leon, *Battle Cry*, 113
U.S. Army, 2, 3; and Al Anbar, 98, 99; and *Battlezone*, 73–74; *The Counterinsurgency Field Manual*, 93–94, 97, 101, 109, 117, 119, 121; and Fallujah, 67, 68; and V Corps, 2, 12; and *Full Spectrum Warrior*, 82; Future Holistic Training Environment Live Synthetic, 160–61; and Institute for Creative Technologies, 76; and Iraqi invasion, 14; Medical Command, 109; National Simulation Center, 161; and VIPE Holodeck, 161

U.S. Marine Corps: *2012 U.S. Marine Corps Science & Technology Strategic Plan*, 141; Air-Ground Center, Twentynine Palms, California, 5; *Among the People*, 98; Center for Advanced Operational Culture Learning, 98; Combined Action Platoon, 114–15; Commandant's Reading List, 98, 114, 129, 133; Computer War Game Assessment Group, 76–77; *The Counterinsurgency Field Manual*, 93–94, 97, 101, 104, 109, 117, 119, 121; Division of Public Affairs, 39; Expeditionary Force Battle Simulation Center, 76; First Battalion, Eighth Regiment, 5, 65; First Battalion, Seventh Regiment, 6; First Division, 2; I Marine Expeditionary Force, 12, 20–25, 68, 104, 105, 157; First Marine Division, 99, 102; Fourth Marine Brigade, 149; History Division Oral History Project, 22; Infantry Immersion Trainers, 76, 161; Infantry Training Battalion, 5; Marine Air Ground Task Force (MAGTF), 15, 21–22, 24, 125; Marine Attack Squadron 211, 23; Marine Attack Squadron 214, 24; Marine Corps University, Center for Advanced Operational and Cultural Learning, 119–20; Marine Corps University Press, 98, 113; *Marine Corps Vision and Strategy, 2025*, 118; Marine Rifleman's Creed, 116; Martial Arts Program, 130; Modeling and Simulation Management Office, 66, 76–77; National Museum, 149–51, 175; Oral History Project, 24; Professional Reading Program, 114; Regimental Combat Team 1, 67; "The Resilience Research Project," 119–20; Rifle Integration Facility, 76; Second Division, 2, 107; *The Small Wars Manual*, 11, 99, 101–2, 106, 109, 119; Third Aircraft Wing, 20–25; Third Battalion, Fifth Regiment, 12, 96; Third Battalion, First Regiment, 31, 51, 62, 64, 65, 66, 69, 78, 107, 108, 109, 111, 112; Third Battalion, Second Regiment, 97, 100; Third Marine Aircraft Wing, 33, 68; Unmanned Aerial Vehicle Squadron 1, 68; Warfighting Laboratory, 88; *Wargames Catalog*, 77
"U.S. Marine Kills Iraqi" (video), 56
U.S. Navy, 20, 40, 73
U.S. Supreme Court, 70

Vest, Kevin S., 23, 24–25
Veterans Empathy Project, 201
video-game makers, 65, 75–76
video games: and behaviors in response to images, 84; and bodily register, 86; and cognitivism, 82–83; as creative expression vs. activity or play, 69–70; and "D-Day Fallujah" television program, 63–64; and decision-making skills, 77; and defense industry, 75–76, 82; development of, 73; effectiveness of, 86; and facial expressions, 80, 81; first-person shooter, 7, 65, 73, 75, 77, 81, 84, 86, 156, 158, 160; goal-directed, 82; Hollywood's influence on, 79–82; and interactivity, 82, 84–85; as killing enablers, 44; as leading to violence, 70; and memories, 83–84; money spent on, 79; and motivation, 82; and post-traumatic stress, 175; and preparation for battle of Fallujah, 69; and presence, 94–95; and procedural memory, 82, 85, 86, 90; realism of, 189–90n64; and simulation, 70–71, 82, 86; and sports, 86; third-person shooter, 66, 94; and urban combat, 64, 65, 77, 82, 87–88, 158; and virtual veterans, 83–84; and war porn, 56
videos, 42; of beheadings, 166; comments on, 49–57, 64; contemporaneity and authenticity of, 57; dehumanization in, 52–53; of fire-fights, 8; killing in, 51, 52; motivation by, 53; taken by marines, 8, 165–66; violence in, 52, 53, 54–55, 56–57. *See also* YouTube
Vietnamese people, 114
Vietnam War, 21, 40, 45, 63, 67, 131, 153; and Card, 72; combat films about, 57; and counterinsurgency, 114; defeatism after, 152; and film, 47–48, 156; and mock villages, 156; nonfiring rate in, 146
violence, 37, 151; and culture, 1; desensitization to, 44; in *First to Fight*, 93; imaging of, 8; Marines' capacity for, 108; and masculinity, 1; of military occupation, 108; and optics of combat, 3; and seeing, 7; and sight, 158; techniques of, 141; video games as leading to, 70; in videos, 52, 53, 54–55, 56–57; and war porn, 42. *See also* death; killing
virtual environment, 160

Virtual Hills and Laguna Beach (television series), 82
Virtual Immersive Portable Environment (VIPE) Holodeck, 161
Virtual Iraq (exposure therapy system), 82
virtual presence, 75–76
virtual reality, 85, 161
V-22 Osprey, 125

Wagner, Richard, "Ride of the Valkyries," 69
Wake Island (1942), 46
Wallace, R. K., 138, 139
Walzer, Michael, 37, 191n6, 197n45
war and combat: aggrandizement of, 37; American intolerance of casualties in, 164; asymmetrical, 101; challenge and mortal danger of, 154; as clean, effective, and gratifying, 152; as competition, 9, 27–28, 69, 86, 101, 145; and culture, 8, 44; digitized, 129; disembodiment of infantry in, 137; drama of, 69; and empathy, 129; enjoyable aspect of, 79; film as idealizing, 156; fog of, 15; as game, 145; immaculate, 14, 179n5; just, 58; literature about, 1, 37–38, 112–18, 170–75; for oil, 51; and photography, 38; as play, 9, 66–67, 69, 70, 72, 75, 104, 145; pleasure in, 8, 42–43, 44, 45, 48, 49, 50, 53, 56, 72; and revelry in fighting, 43; simulating feel of, 156; and sports, 67, 156; structured expectations of, 45; subjective experience and sensibility of, 56; symmetrical, 101; synthetic views of, 158–68, 167; three-block, 89, 96, 97, 103, 120; in videos, 53; visualization of, 5, 6. *See also* urban combat
Warhammer 40,000 franchise, 131–32
War on Terror, 116
war pornography, 7–8, 49, 81, 86, 146, 164–66, 183n12, 202; defined, 40–41; and empathy, 100; as force projection, 8; on hard-core to soft-core continuum, 50; by Iraqi insurgents, 110; Islamic State, 166; and killing, 160; music video–like, 50; production of, 50; as shown in Marine boot camp, 45; and Swofford, 171; and YouTube, 40, 44–45, 70, 160, 202. *See also* videos; YouTube

warriors, 153, 155; and battle robots, 145; ethos of, 117, 128, 145, 159; and film, 1; ideal of, 113, 115, 116, 117–18, 127, 147; mental attitude of, 8, 167; and self-possession, 113, 117, 159, 167; as shaped by weapons, 131; subjectivity of, 159–60, 167, 170, 171, 173–74, 176
War Stories with Oliver North (2001–10; television series), 50, 59
Wayne, John, 47, 51, 81
weaponry. *See* automated weaponry; high-technology weapons systems; remote weaponry
Webb, James, *Fields of Fire*, 113
Wedemeyer, J. Scott, 22
Weigley, Russell, *The American Way of War*, 178–79n4
Welsh, Mark, III, 163
West, Bing, 6–7, 25, 28, 29–30, 68, 101, 106–7, 109; *The March Up*, 157–58; *The Village*, 113, 114–15
What Price Glory (1926), 39
"Which Way Would You Run?" (USMC commercial), 124–25
Wii, 85
WilcoUSMC (video uploader), 54
Williams, Lloyd, 149

Williams, Raymond, 108
Wire, The (television program), 59
Wolfenstein (video game), 69
Woltman, Clyde, 22
women, 136–40; combat roles for, 137–38, 140, 142, 154–55, 156; and sexual reproduction, 154–55; violence against, 42. *See also* gender
World War I, 38, 47, 146
World War II, 21, 38; air power in, 13; and carpet bombing, 19; dehumanization of enemy in, 52; films about, 46–47, 57; and History Channel, 61; and Hollywood combat movies, 43; and live-fire courses, 156; non-firing rate in, 3, 4, 146
Wright, Evan, *Generation Kill*, 42–43, 59, 62, 154
Wynn, Mike, 100

Yemen, 1, 142
YouTube, 42, 65, 160, 164–66, 202; commentary on, 49, 50–52, 53, 54, 55–57; and marines, 7, 202; and masculinism, 7–8; and war porn, 40, 44–45. *See also* videos; war pornography

Zinni, Anthony, vii–viii, x